SCIENCE AS A COMMODITY

Science as a Commodity:
Threats to the
Open Community of Scholars

edited by

Michael Gibbons and Björn Wittrock

Longman

Longman Group Limited,
6th Floor, Westgate House, Harlow, Essex CM20 1NE, UK

First published 1985

British Library Cataloging in Publication Data

Science as a commodity.
  1. Research—Social aspects
  I. Gibbons, Michael II. Wittrock, Björn
  306'.45   Q180.55.S62

ISBN 0–582–90204–5

Typeset in 10/12pt Linotron Bembo by
The Word Factory, Rossendale, Lancashire
Printed in Great Britain by
Butler and Tanner Ltd, Frome, Somerset

Contents

Foreword

This collection of essays originates from an international conference held in Stockholm on 2 and 3 May, 1983. The meeting was organized by the Swedish Council for Planning and Coordination of Research (FRN) through its Committee for Future-Oriented Research.

The council has its place alongside the councils for natural sciences, medicine, and social and humanistic studies. Its special obligation is among other things, to initiate and support work which transcends the boundaries of the sectoral system, to stimulate a dialogue between the academic community and society at large, and to develop international links in areas not covered by the other bodies. The council's Committee for Future-Oriented Research has in its turn as its main task the initiation and support of studies of a fundamental nature which will increase our understanding of long-term changes in society and the environment, and lead to comprehensive knowledge which could be used to shape the future.

Since scientific knowledge is of the utmost importance, both as a resource and as an agent for change in society, the tasks just mentioned make it essential to promote the self awareness of scientists and science policy makers by making science itself the object of study and of public debate. The conference reported in this volume reflects this outlook. Authors were asked to make a critical evaluation of internal and external developments of the scientific enterprise, viewed in a broad international and historical perspective. The topic of the conference was intentionally defined in a provocative way. The same formula has been retained for the title of this book. It stimulated a very candid debate.

The concept of the conference was worked out initially by a small preparatory group including Uno Svedin, Roger Svensson and Björn

Wittrock. The group was also responsible for the selection of the contributors and other participants. In practical matters Berit Örenewall provided invaluable assistance during the preparatory phase and at the meeting itself. The two editors who were invited to contribute to this volume, Michael Gibbons of Manchester University, and Björn Wittrock of Stockholm University, consider at length the relevance of the chosen topic and summarize both the participants' and their own reactions to it in the Introduction and the Postcript.

The sponsoring organizations wish to thank the preparatory group, the editors, the authors of essays and all participants of the conference for their generous contributions. Thanks are also due to the Wenner-Gren Centre and Stockholm University for providing a splendid environment for the meeting.

Torsten Hågerstrand
Chairman of the Committee for Future-Oriented
Research

Lund
August 1984

Introduction

Michael Gibbons

An introduction to a collection of essays is meant, at the very least, to indicate what the collection is about. It may even attempt to give prospective readers some sort of map to guide them to specific contributions which may be of more individual interest. It is not uncommon, therefore, for the editors of collections such as these to claim that, despite the different perspectives of the authors, the essays 'hang together' in some way which it is the purpose of the introduction to make clear. The coherence perceived by the editors provides both a broad description of the problem and a large-scale map providing routes to more specific issues. Alas, in the present case, the editors are able to make no such claim nor are they able to provide much in the way of a map. It is incumbent upon them, therefore, to say why this is so.

The difficulty lies primarily in trying to formulate the problem. What precisely is the problem that is described by the title, 'Science as a Commodity: threats to the open community of scholars'? Several subsidiary questions arise: 'Do the essays describe an historical development or a new situation for the open community of scholars?' 'Are they about the current crisis in the universities regarded here as the institutional bases of research and scholarship in the twentieth century, or are they about the current performance of the academic community?' 'Are they about the invasion of the universities by other social institutions: industry and government generally, the military, or public interest groups?' 'Are they about a critical shift in the relations between science and society?' The short answer is that reference to each of these questions can be found somewhere in this volume. The difficulty is that the questions point to a deep-seated problem that is more deeply felt than it is understood. It was in the hope of getting a little clarification of some of the issues involved that the Swedish Council for Research and Planning invited senior academics, industrialists and civil servants to meet for two days in the Wenner-Gren Centre in Stockholm in May 1983. It was not a large gathering but both the papers given and the discussion that ensured exhibited an intensity of seriousness that is not often found in international meetings concerned with similar problems. The reason for this, the editors believe, lies in the fact that, as formulated, the title of the workshop – Science as a Commodity: threats to the open community of scholars – touches a deeply held belief of a large number of people carrying out different tasks in a wide variety of institutions broadly concerned with the development of science and technology. That belief is exposed in the tension between the way science (or knowledge) is *used* in our societies and the way in which it is supposed to be *generated*. The tension arises because it is not clear whether the knowledge that is generated is being used properly or whether if it were to be generated properly it would be usable.

THE LONG PERSPECTIVE

As has already been indicated, in its most general formulation, these essays are about a changing relationship between science and society; or, perhaps more accurately, between the open community of scholars – to the degree that it survives in contemporary institutions of higher education – and society. What is more or less deeply felt is that something is wrong but precisely what is harder to express clearly. As the quotation from Peter Scott at the opening of the paper by Michael Gibbons puts it, we are in difficulty because we do not possess an appropriate language to deal with the problem, but we know that what is wanted is a language that is more than technical or administrative, and which 'can impose a moral structure on our exploding experience'. To develop, or more properly, to discover a new language is undoubtedly a complex activity but no sensitive reader of these essays can fail to appreciate that Scott has indentified a critical issue.

Still, a new language, if it is to develop can only emerge from experience; and experience, in turn, is channelled or guided by images. Perhaps the dominant images governing current thinking about science in its multiple relationships with society are those associated with the biological notions of growth and development. Science, in particular, during the last one hundred years has not only exhibited exponential growth but also has moved from the periphery to the centre of the social, economic and political life of most advanced industrial societies. Such growth and such a transformation of the role of science have left neither science nor society unaltered. To pursue the analogy further in biological terms, it could be said that the growth of science which has to a large extent been unplanned has altered the traditional relationship between its form and its function. In biological organisms form and function evolve in relationship to one another. As the biologist D'Arcy Thompson observed at the turn of the century in his studies of the growth of organisms, there is a tendency for bodily *surface* to keep pace with *volume* through some alteration of its form. In fact, he argues, a great deal of evolution is involved in keeping a balance between the volume of an organism and its surface as growth goes on. Obviously, the rapid growth of higher education in so many western nations cannot be likened precisely to the growth of a living organism but it draws our attention to the important relationship between the inside of an institution (science or higher education in this case) and its outside (the institutions that comprise the rest of society). The point is that as the volume of science has grown so its surface – that is, its relationship with the rest of society – as had to alter. And, in line with the biological analogy, this growth should bring with it an alteration of both the form and the function of science. The organism is clearly threatened if growth advances so quickly that the relationship between form and function cannot be maintained. It is arguable that this is the situation not only of science but of many of the best contemporary institutions of higher education as well. Both science and the universities have begun to experience a condition of functional overload; of trying to do too many things without altering the form of the institution.

If the image of the growth of a cell is helpful, still it is not explanatory; for what is needed is an understanding of what has taken place at the core of the cell throughout its period of rapid growth. A possible explanation is that through its history, but particularly during the last hundred years, science has abandoned its search for true, certain knowledge of causal necessity and become a commodity. The notion of science as a commodity raises a number of interesting and difficult questions, but let us consider just two for the moment. Firstly, can the contemporary production of scientific theories, methods and data be validly compared to a manufacturing process whereby goods are produced for exchange and, therefore, intended to be purchased either in their own right or traded for some other commodity? And, secondly, do scientists really function as labourers in the production of scientific knowledge in a way analogous to that experienced by workers who operate the machines that provide the endless streams of goods and services that make up our standard of living?

With regard to the first, it is noted in several essays that science no longer strives to attain the ideal of discovering true certain knowledge of causal laws which motivated firstly Greek science and, much later, the founders of the scientific revolution of the sixteenth and seventeenth century of western Europe. On the contrary, current science is hypothetical knowledge and its conclusions therefore are only provisional. To the extent, then, that science is hypothetical it is about making models that more or less adequately explain natural phenomena. To the extent that these models are open to revision, they are cut loose from the need to try to grasp reality within a single view. As a consequence, specialism has its chance to develop. These two aspects of contemporary science provide the basis for the extension of modelling methods to a wide range of phenomena and make it possible (but no more than that) for scientific technique to be applied in a number of different contexts and be exchanged for rewards beyond those of purely professional recognition. That scientific techniques actually come to be so applied and exchanged in this way perhaps owes less to developments internal to science itself, and more to the diffusion throughout western societies of a mode of social organization based upon production for exchange. On this view, it is hardly surprising that, in an industrialized society, science should be treated more as a commodity.

The second question is concerned less with the epistomological aspects of science than with the activities of scientists engaged in producing theories, models and data. That scientists could validly be regarded as labourers in the field of knowledge might, at first sight, seem rather farfetched. The scientific community, after all, has always placed the highest priority on, and given its most prestigious rewards to, those individuals who manifest originality and creativity. Yet, it is but a short step from Thomas Kuhn's dichotomatization of science into revolutionary and normal science to the realization that a large part (perhaps the largest part) of scientific research has become highly routine and that, for many, originality and creativity, such as it is, take place in the context of a group research project. So pervasive has been the development of team research that it is at least worth while asking about the extent to which the older ideal of individual originality actually survives.

That science in our culture might have become a commodity is, then, a notion that deserves some serious thought. It is of particular concern for those involved in higher education, of course, because both in their own self-image and in that of society more broadly, the universities are still believed to be an open community of scholars. Science that is practised outside of this context may well have taken on some of the characteristics of industrial production but, surely, in the universities the values of the open community of scholars still survive. It seems that the possibility that things might be otherwise has diffused widely and the growing awareness of this lies at the root of the belief, described above as being more widely felt than understood, that the universities are under threat. Yet in the universities, too, the provisional nature of all knowledge has led on the one hand to an emphasis within science upon technique and, on the other, as techniques have proliferated, to the fragmentation of the field of knowledge into tiny areas which are intensively cultivated. The result is the production, even in the universities, of professional specialists who may be more accurately described as researchers than scientists. The question, then, of whether the notion of an open community of scholars is anything other than a high form of nostalgia is of crucial importance for the future of higher education. But what, precisely, would be lost to science if it were clearly and widely recognized as a commodity, and scientists were regarded as knowledge producers remains somewhat obscure. The same, however, cannot be said of the impact of this change on society. For at the root of this change lies the possibility (one might say the fear) that, as a commodity, science will be manipulated for their own ends by those who allocate resources for research. Given the economic and military potential of contemporary scientific knowledge, such manipulation cannot be regarded with equanimity.

SCIENCE IN TRANSITION: A NEW PATTERN OF INTERESTS

There is a real and urgent sense that the scientific enterprise is in transition. Several papers draw attention to this fact by referring to the idea of a previous golden age of science (Helga Nowotny). Sometimes the purpose of this reference to a golden age is to mark its passing (Michael Gibbons) or to point to its naivety (Emma Rothschild) or to contrast it with the emergence of science as useful and, therefore, of interest to the powerful (J.-J. Salomon). Nonetheless, though several authors formulate a view in relation to some real or imagined previous age, there is an underlying impression that with the advance of the scientific revolution something was definitely gained but also something was lost and that in general both science and society are somewhat worse off for it.

For example, Sheldon Rothblatt, in exploring how the notion of an open community of scholars has developed historically, discusses the existence of two distinct sets of pressures operating on the scientific community which have served to define what openness could mean. 'The first set of pressures consists of demands for a wide range of scientific services for government, industry and the military . . . The second set derives from the internal constitution of science, from its cultural or value system and from the institutions that scientists have built or have cooperated in building in order to maximise the conditions under which their work is performed.' Here, the operation of these two pressures, an external pressure attempting to mould science to social need and an internal pressure aimed at acquiring social acceptance for scientific knowledge, resulted in the professionalization of science which has given the scientific enterprise the shape it has today. Professionalization is itself a manifestation of the existence, at a deeper social level, of a principle of exchange which, it is often said, is the hallmark of an industrial society. In the case of science, professionalization was a way in which groups with specialist expertise were able to obtain a degree of autonomy in return for delivering some other social good (Gibbons, Rothblatt). This is an important development, not least because in adopting professionalization as the route to social legitimation, scientists (or more specifically the scientific leadership of the day) gave tacit acceptance to the validity of the principle of exchanging knowledge for something else. Perhaps no other option was possible; perhaps all societies achieve cohesion by means of a more or less complex network of exchange relationships. But, it cannot be denied that in an industrial society such exchange relationships have come to be very narrowly interpreted because of the general presence of an experience with the production and sale of commodities. As many scholars have already noted, the march of industrialization is, in some respects, the march of commoditization – in the sense of the exchange of goods and labour for one another. In such a society can knowledge producers expect to be treated differently?

It is not only in its external relations that science has tended to become like any other commodity. Implicit in this development has been the emergence within science itself of project-oriented research and its concomitant, the research team (Nowotny). The birth of this social form of research organization has its origins in the Manhattan Project: that huge multidisciplinary scientific and technologial activity that produced the first atomic bombs. Helga Nowotny argues that this type of project work is now the principal mode of carrying out scientific research. Her case is amply supported by Hermann Grimmeiss who, in trying to explain the technological requirements underlying international competition in the semiconductor industry, stresses the importance of both multidisciplinarity *and* intensive specialization in product development. Hermann Grimmeiss is making the point that scientific research is not an option for firms at the forefront of technological development, it is an imperative. And, insofar as the advanced industrialized nations come to see that their principal economic advantage lies in their ability to utilize knowledge resources, one may expect more and more significant scientific advances to occur (as has been demonstrated repeatedly in the case of the Bell Telephone laboratories in the USA) in the context of industry. It seems incontrovertible that if this were to occur more widely both science and the open community of scholars would have been transformed out of all recognition.

INSTITUTIONS AND INDIVIDUALS RE-EXAMINED

There is abundant evidence in the papers by Emma Rothschild, Jean-Jacques Salomon and Dorothy Zinberg, that this transformation is well under way; so much so that both science and the universities are now seen to be operating in a new context. What, then, is novel in this new situation? Firstly, there is the pervasive growth, particularly in the USA, of military research and development. The threat here, as Emma Rothschild argues, arises from a gradual encroachment on the ability of scientists to exchange information freely. The line between military and non-military research activities is clearly a hard one to draw and that is why encroachment has been very gradual. In some areas of research – electronics for example – military research is clearly an important source of revenues for universities but it can also divide them into 'go' and 'no-go' areas and, *a fortiori*, it can divide academic communication into those topics which can be fully discussed and those which cannot be. Of course, because military research is carried out in secret, it is difficult to determine just where the borderline is since, by definition, no questions can be answered about it. Such research, Emma Rothschild argues, surely threatens the continued viability of an open community of scholars in the universities.

Secondly, the rules of the game facing those universities which are trying to gain a measure of financial stability through research grants and contracts are changing dramatically, at least if the American experience is anything to go by. The situation, there, in relation to industrial research grants is clearly described by Dorothy Zinberg in her paper. One of the points made in the paper is that university – industry relations of the type embodied for example between Hoechst AG and the Massachusetts General Hospital, or between the White-head Institute and the Massachusetts Institute of Technology are so demanding that they may alter, in a substantive way, not only what it means to be an academic researcher, but also the balance of the curriculum offered by the university. Still, the universities are in this position in part because governments generally have become sceptical of the willingness of universities to change themselves and have begun to trim their budgets accordingly. There is little option for universities who wish to pursue excellence in research but to seek alternative funding where they are able. This involves, necessarily, a more careful attention not only to customer needs but also to the investment of resources in marketing the research skills of the university. Needless to say, these jobs are not done by the researchers themselves and so, even in the university, the traditional right of faculty to pursue their own research interests is being gradually eroded to maintain the research reputation of the institution.

Helga Nowotny observes that it is time to abandon the assumption of the golden age of science; that it is sufficient if 'good men do good science'. As it is currently organized, the performance of good science is divorced from the wider political (and moral) dimensions of life and, therefore, 'scientific openness, in the traditional sense, is bound to fade out'. What, then, can be done? Somewhat pessimistically she concludes, 'Good science and good men are insufficient. Good women have had no chance, as yet to alter the rules of the game . . . (an) appeal, even a strong moral appeal to individual responsibility will not do either'.

Similar concern about the ability of individuals to transform the current situation of both science and the institutions of higher education appear in several places in this volume. In keeping with the *Zeitgeist* of our times, we seem to want to pin our faith on better organizations and more flexible institutional relationships. But if men and women individually cannot be expected to reach the degree of moral responsibility necessary to prevent the final dissolution of the open community of scholars what, concretely, can we expect from an organizational revolution however profound? How can institutions be designed so that they will promote good science and eschew the temptation of pursuing problems that are merely 'technically sweet'? 'Or would we,' as Nowotny puts it, 'opt for less?'

In his paper Jean-Jacques Salomon observes that one hundred years ago the question of whether science was a commodity would have been regarded as an aberration and would have received scant attention. Fifty years ago, the same question would have been pro-vocative; that is, it would have been worth discussing just long enough to discredit it. Today, the status of science as a commodity is more or less taken for granted. At least, none of the essays in this volume offer any dissent from that formulation. What, then, *has* happened to science during the last hundred years or so? What *is* happening to science now? Any coherence of this volume derives from the fact that each author is attempting to deal with one or other of these questions. Given the complexity of the current situation which has been adumbrated, albeit briefly in this introduction, it should come as no surprise to anyone that the essays themselves draw upon different data and reach different conclusions. This, at bottom, is the reason why the editors have felt it unwise to claim that they have discovered a unity of perspective which the authors have neither sought nor desired.

PART ONE
The Long Perspective

The Changing Role of the Academic Research System

Michael Gibbons

'Both modern society and higher education are struggling in ways that are incestuously linked with equally indifferent success to establish a meta-language that is more than technical and administrative and which can impose a moral structure on their exploding experiences.'
Peter Scott, *Times Higher Education Supplement*, 27 August 1982.[1]

'How do you explain school to higher intelligence?'
Elliot Taylor in *E.T. – the extra-terrestrial*, 1983.

The juxtaposition of the title of this paper, 'The Changing Role of the Academic Research System' with the notion of science as a commodity and the threats this poses of the open community of scholars is extremely suggestive for anyone who is concerned with the situation in which the universities find themselves today. That situation, briefly, is one in which the whole relationship of the university to society is being called into question and the curious thing is that despite all the words both spoken and written on the topic there is a growing awareness that we no longer possess an adequate language with which to discuss the role of knowledge in our economic, political and cultural life. As Hannah Arendt has pointed out, when our basic concepts fail it is because we are stuck in the gap between the end of one tradition and the beginning of another.

Let us consider some of these concepts for they contain considerable symbolic potency: science, commodity, open community, scholars, role, academic, research system and threat. There is hardly a word in this collection which is not ambiguous and about the meaning of which there has not already been produced enough books to fill a small library. Much of this writing, however, gets its focus from the tension that is created by putting the words into a specific combination. The tension arises because of a possible incompability between science and commodities; between scholars and academics; between communities and research systems; between institutions which are open and those which are functional. In other words, there is a sense of opposition between the scientific enterprise as an open community of scholars and the knowledge production enterprise as an academic research system. Futher, one cannot avoid feeling that the notion of the 'threat' arises because the former which represents the ideals and aspirations of an earlier tradition is about to be replaced by the latter which will, in its turn, found a tradition of its own. But, as long as the new tradition is not yet firmly established, there is bound to be a sense of fear that something which was once valued is now passing away. The question, therefore, that I would like to address is whether the concern about science and the values of the open community of scholars is anachronistic, that is, out of time with the age in which we are living. Hannah Arendt once observed that, 'the beginning and the end of a tradition have this in common: that the elementary problems of politics never come as clearly to light in their simple and immediate urgency as when they are first formulated and when they receive their final challange.' She goes on to quote Jacob Burckhardt's evocative metaphor that the experience which constitutes the beginning of our tradition is like a fundamental chord which sounds its endless modulations through the whole history of western thought.

> 'Only beginning and end are, so to speak, pure or unmodulated; and the fundamental chord therefore never strikes its listeners more forcefully and more beautifully than when it first sends its harmonising sound into the world and never more irritatingly and jarringly than when it still continues to be heard in a world whose sounds – and thought – it can no longer bring into harmony.'[2]

Therefore, it is necessary to ask whether the modulations of that fundamental chord which reverberate every time we speak of science and of its progress through the mediation of an open community of scholars is any longer able to bring harmony into the contemporary relationship between science and society.

THE ACADEMIC RESEARCH SYSTEM

The academic research system may be used as a shorthand expression to convey the situation of science and scientists, natural, social and medical in a contemporary university setting. As I shall try to demonstrate, this system is the carrier of the contemporary scientific project.

It is important, therefore, to grasp what its essential characteristics are. There are several ways in which this could be done but one of the more suggestive, to my mind, derives from a comparison between what I shall refer to as the Aristotelian ideal of science and the contemporary ideal of science. To compare these two ideals is particularly useful, here, because the original idea of a 'scientific' project organized around an open community of scholars derives from the Greeks, and not, as we shall see later, from Galileo, Bacon or the others who launched the scientific revolution of the sixteenth and seventeenth centuries. Contemporary research requires a different form of organization, and the comparison we are proposing to make will throw some light upon contemporary scientific practice. This is a minimum requirement if we are to make any judgements about whether this project is under some sort of threat.

It may be helpful to clarify what I mean if I briefly set out the main elements of the Aristotelian ideal of science and compare them with those that appear in the contemporary ideal. For Aristotle and for those who adhered to the classical system of Greek philosophy, science was true, certain knowledge of causal necessity. Its aim was to contemplate the unchanging laws of being as revealed by intuition and, subsequently, expressed in the first principles of a metaphysics. Its method was based on the syllogism which moved from self-evident first principles, deductively to truths about the celestial, the terrestrial, the moral and the political orders. Science was a closed system of interlocking propositions and deductions which were dependent upon prior metaphysical assumptions. Because it was a closed system – a logical complex to which one could revert should the system be attacked from any quarter – it could be mastered by a single individual.[3]

Now, it is worth spending a little time examining the classical idea because in almost every respect the contemporary ideal of science diverges from it. Contemporary science is not true but is on its way towards truth; its conclusions are not certain but only probable; its theories are subject to constant revision. Contemporary science does not aim at knowledge but at formulating hypotheses, building models, erecting theories and systems. Modern science is not concerned with causes such as end, agent, matter and form, but with establishing correlations to be used as a basis for the complete explanation of all data in terms of networks of mutually intelligible relations. Contemporary science is not necessary but rather is content with verified possibility – not with what must be the case but with what, in fact, is so. Whereas classical science made use of logic and its principle technique, the syllogism, modern science is concerned with empirical method, with gathering data, with the formulation of hypotheses, with the verification or falsification of them under precisely controlled conditions. Modern science is not an individual achievement but is the result of group collaboration. Its aim is not contemplation of the unchanging order of Being, but utility: contemporary science aims to have practical results.

Now, there are many important implications which follow from this shift from the classical to the contemporary ideal of science. For our present purposes, however, it will be sufficient to examine the philosophical, social, methodological and practical aspects of this transformation. When this is done it becomes possible to see in a clearer light how science might have come to function as a commodity.

The Philosophical Dimension

Contemporary science no longer aims at certain knowledge but is satisfied with verified probability. Thus, the contemporary scientist knows that his conclusions are at best probable and the best evidence that he has for this is based upon his personal experience that theories are subject to periodic revision. New scientific text books are necessary because new theories do not simply build on existing ones and thereby increase what has to be learnt but in many cases new theories replace existing ones altogether and thereby alter the way in which one has to think about and investigate a problem. Compare, for example, books on chemistry before and after Dalton, on physics before and after Heisenberg, on geology before and after Tuzo Wilson, on biology before and after Watson and Crick.

What the history of science has been disclosing is that the scientific project is about *understanding* nature, not knowing it in the sense that Aristotle had in mind, and that the number of ways in which the material universe can be understood is very large indeed. It is now more clearly appreciated that the range of models, theories, and systems which we can construct so as to understand our world is limited only by the power of our imaginations and, some would say, the range of cultural resources available. With so large a range of possibilities to choose from, the problem of why one or other model is chosen becomes an interesting historical question and leads one to consider whether the choice is based entirely on purely cognitive considerations.

To admit that theories might be chosen and developed vigorously not only for their explanatory potential but also because they may advance an individual research career or keep a government in power or promote industrial competitiveness opens the door to the possibility that the value of scientific knowledge is relative to its function in some larger system such as the military/industrial complex. Nonetheless, if one is unable to explain adequately why one theory is chosen rather than another, then the conclusion may be drawn that the choice is purely arbitrary and this threatens to undermine both respect for the conclusions of science and the legitimate authority of the scientist himself.

Aristotle was always very careful to distinguish between science and opinion; the former being concerned with what was true, eternal and changeless, the latter being concerned with what was changeable and corruptible. We do not make these distinctions. Instead, we seek the 'best available scientific opinion of the day' and so acknowledge the provisional nature of all scientific findings.

There is, at the present time, a great confusion about the nature of scientific knowledge and its solution is a philosphical question which unfortunately is not of very great interest to those who practise science. But, as long as the question is unresolved, there is always present the danger that 'the best scientific opinion of the day' will become 'merely another opinion' in the vast array of views that can be sought or bought on any matter of social relevance. Without a clear view of what kind of knowledge is produced by the scientific enterprise and the grounds of its validity, the scientist, unable to claim any greater right to be heard than anyone else, may find his role obscured in a battle among experts.

The Social Dimension

For Aristotle, science was a fairly compact affair. That is to say, it was a closed system and access to its first principles was obtained by dint of the hard work of the pupil studying the works of the master. Science was an individual achievement, and excellence was judged according to the ability of the student to defend or elaborate upon the teaching of the master in relation to some more or less complex problem, without changing the first principles.

By contrast, the carrier of science today is the research team and that team is a social entity. At the simplest level we can observe that no individual can know the whole of mathematics or physics or chemistry or biology. Such knowledge is no longer possessed singly by an individual but collectively by the members of a group. Further, if one wants to practise as a scientist one must first of all become a member of one of these research groups.

Entry to a research group is not automatic. There is much to be learned and the style of learning is that of apprenticeship – and a long apprenticeship at that. So it is that most scientists have passed successively from primary school to secondary school, from university to postgraduate studies, and through the initiation ritual of a PhD. During this lengthy period, scientists become familiar with the specialist language of their field. They learn the correct techniques and procedures to be employed in solving problems. They become masters of the conceptual schemes, models, theories which guide their thinking in respective fields. They join the appropriate associations, attend conferences, read journals, contribute to the publications and design the tools and equipment they may need. The upshot of this is that science is a specialism, that scientists are specialists and, it must never be forgotten, that the aim of these specialists is to keep their specialism alive and flourishing.

So, in the research group we observe what has come to be called a social definition of knowledge. Within each specialist group problems are identified, techniques specified, solutions tendered and criteria of the appropriateness of the former to the latter decided. All of this takes place within the group. Since, with the collapse of the Aristotelian ideal, there is no longer any external or abstract reference against which the 'truth' may be measured, discussions about what shall count as verified knowledge are reached through the operation of the peer-review system; that is, through the collective judgement of one's colleagues. In contemporary science, knowledge is *consensual*. Knowledge maintains its scientific status only so long as that consensus can be supported. It is perhaps due to reasons of group dynamics that science changes so slowly. It was the slowness of the pace of scientific change that gave the eminent scientist Max Planck reason to complain that a new scientific truth does not triumph by convincing its opponents and making them see the light but rather because its opponents eventually die, and a new generation grows up that is familiar with it.

The shift from science as an individual achievement to a group or team activity carries with it the first intimation of what has come to be called the academic research system. Further, if the coherence of the system is to be maintained, norms must be elaborated to direct behaviour appropriately. To the integrity and authenticity of the individual scientist in relation to his work there has been added a further set of social guidelines designed to maintain the solidarity of the specialism and to ensure that its activities are clearly distinguished both from other specialisms and especially from other groups who seek to legitimate their ideas and activities by prefixing them with the adjective 'scientific'. Some of the norms which provide these guidelines are, by now, well known and include such things as universalism, organized scepticism, communality and disinterestedness.[4]

Specialization in the sciences and the establishment of norms to control behaviour within the scientific community form part of the larger process commonly referred to as the institutionalization of science. The institutionalization of science took place in Western Europe throughout the nineteenth century and, during this century, science gradually came to be recognized and accepted by the other major social interests and institutions – the Church, the State and Industry. However, it was not until the middle of the twentieth century that science was able to move from the periphery to the centre of political economic life of the majority of the countries of Western Europe.

For this latter development to occur, the nascent scientific community had to establish closer and firmer links between itself and the rest of society. To accomplish this it followed a programme which many other groups were experimenting with – the scientific community sought to join the ranks of the professionals. Throughout the late nineteenth and early twentieth centuries, the transformation of science as an individual achievement to science as a fully professional activity was accomplished, but, as we shall see, to become a profession involves certain reciprocal relationships between a profession and the society which sustains it, in the sense of recognizing the activity as valid and useful and granting it the necessary means to survive. 'A profession is a socially rooted and supported vocational enterprise of full time practitioners who earn their living by providing a vital social service through the utilization of expert and esoteric skills.'[5] In other words, any profession (scientific ones included) exist in a kind of exchange relationship with society – social support is given for services rendered. As far as the scientists in the late nineteenth century were concerned, the social support they needed was more than good will and social acceptance. These were important but to some extent they had already been achieved. What was required now was something much more concrete – money.

Money – what today we would call resources – was needed because professionalization depends upon large-scale societal investments and public support of scientific institutions.

'Years of traning in the universities, the establishment of postdoctoral and other research facilities, and the development of manpower programmes all depend on resources allocated through public policies that foster professional objectives.'[6]

In other words, the apparatus of specialization is expensive, so expensive in fact that it is only through continual government support that the security of the scientific project itself can be guaranteed. That, at least, was what the early advocates of professionalization believed. It was clear to them, even in the late nineteenth century, that neither the charitable foundations nor industry could be relied upon to support scientific activities systematically and at the appropriate levels. And so the campaign to solicit government support began.[7]

There are several characteristics of professional activities to which we should refer. Firstly, a profession is a socially rooted and supported vocational enterprise of full-time practitioners who can earn their living by providing a vital service through the utilization of expert and esoteric skills. Secondly, in order to be able to achieve the appropriate levels of expertise, the professions are granted a good deal of autonomy over their internal affairs: setting standards, training new specialists, invoking disciplinary powers, and establishing and enforcing professional codes of conduct. Thirdly, this autonomy is based upon a recognition that a profession operates in a realm of expertize which those outside can make no claim to and which they, at best, understand only very generally. Fourthly, for most professionals, therefore, a considerable degree of trust is required if the profession is to perform its work satisfactorily. That trust can break down if the public, generally, or some important social group begins to suspect that the profession is not living up to its ideals. A reciprocal obligation is, therefore, placed on the profession; in return for support, it is to fulfil its professional responsibilities. Fifthly, when this trust begins to break down, one is likely to find that an imbalance of power develops and the heretofore accepted autonomy of the profession is interfered with or some attempt made to constrain it.

I have stressed the reciprocity of the relationship between a profession and the society which supports it, firstly, because it is essentially an exchange relationship and, secondly, because it implies that any profession, science included, is necessarily divided in its commitments. With regard to the former, the gradual emergence of specialisms had lead to the piecemeal establishment of a network of exchange relationships between the specialism and the appropriate legitimating institutions, whether it be government, industry or some other groups. This development, which began in the nineteenth century and has continued in the twentieth century, underlies the emergence of what has been called the academic research system. This system, it can now be seen, comprises a more or less loose confederation of specialisms, each of which could be accurately described as a subculture. Within this confederation we can identify both horizontal exchange relationships (those which bind the subcultures to one another) and vertical exchange relationships (those which bind the subcultures to society). The important thing to note is that once the specialism is established, the vertical link may be the stronger, because it provides the channel along which resources flow. A community of scholars on the other hand, seems to stress horizontal relationships; yet it is difficult to be very precise about what the various specialisms share expect perhaps a common commitment to methodological rigour. Any profession must not only obtain the resources necessary to build and maintain its specialism but, because there are personal stakes and careers in research to be considered, there is always the temptation to increase personal influence and, thereby, obtain power. Thus, one finds this professionalized science becoming increasingly involved in political activity to protect its interests as well as to enhance its influence. It is an observable fact that 'as science has attained its present magnitude, conflicts over the allocation of resources to its various sectors, over the determination of priorities and over the accountability of its leaders have become more and more visible and frequent'.[8]

The recognition of this tendency – that as professions develop so they become involved in political activity – is an essential ingredient to understanding the political transformation of the scientific project that has occurred, mainly in the second half of the twentieth century. Moreover, the political transformation of science also calls for a new kind of leadership. In the days before the scientific project was politicized, leadership in the scientific community was awarded to, or assumed by, those who were leaders in their various fields – what might be called paradigmatic leaders. Their authority was rooted in the intellectual contributions which they had made, or were making, to the intellectual development of their subject. Today, we see the emergence of a new type of leadership – institutional leadership. The reciprocal relationship which we have identified as the basis for the successful professionalization of science cannot be sustained by intellectual brilliance alone. Institutional leadership is needed to deal with organizational and political imperatives; to put clearly to government and industry the needs of the enterprise; to orchestrate the consensus both within and outside the research system. It is clear that the qualities required to organize, promote and defend the institutional interests of the academic research system are considerable. Further, it is also clear that scientific achievement alone is an inadequate criteria for institutional leadership. Precisely how this type of leader emerges is far from clear but, whatever the process, institutional leaders have become increasingly important and influential in scientific projects in Europe over the decades that have elapsed since the end of the Second World War.

The Methodological Dimension

There is a further dimension that characterizes the change from the classical to the modern ideal of science. This may be described as the shift from logic to method. The modern scientist places his reliance not on his basic laws or hypotheses but on his method. After all, it was method that brought forward the laws and principles in the first place and it will be method that revises them if and when the time comes for revision. Now, as we have seen, scientific method is not the possession of a single individual but also of the group. Indeed, there is a close relationship between the emergence of specialists groups in science and the successive differentiation of methods. Galileo's discoveries were just one example of a general methlogical breakthrough. The development of science subsequently has been a process of elaboration and refinement of that general method and this has involved a long and painstaking enquiry in search of the relevant data and the appropriate techniques and theories to understand them. Now, while the successive differentiation of methods goes some way to explaining what is happening on the cognitive terrain of the contemporary scientific project, it also has implications for the institutional life of science. Because contemporary science bases its credibility on the validity of its methods of handling and explaining data, a methodological ethic has tended to dominate the value system of science. While this is as it should be when one considers the cognitive aspect of science, the result has been to leave the institutional ethic – that is the social function – of science relatively underdeveloped. The consequence of this is that in its interaction with society, particularly with government and industry, the institutional leaders of science have tended to promote science as an instrument or tool and have given scant attention to the ends to which their energies have been directed. In fact, so successful have been the institutional leaders of science that it is now virtually impossible to think of science in any other way. The idea that scientists only make discoveries while it is up to others to put them into practice has for a very long time not only worked to salve the consciences of many scientists but also to absolve institutional leaders from giving serious thought to the wider social, ethical and political implications of their work. Pursuit of such methodological purity, cognitively, and of the instrumental value of science, institutionally, had led the scientific community to adopt a more or less uniform stance with regard to the prevailing political system. According to Joseph Haberer, since its inception modern science has adopted a tactic of 'prudential acquiescence' to legitimate authority. This tactic, according to Haberer, can be identified not only in the scientific programmes of Galileo and Bacon but in the activities of scientists in the Weimar Republic and in the Manhattan Project.[9] The tactic of prudential acquiescence is currently being tested again in the field of genetics but it is still not clear what the final outcome will be.

Prudential acquiescence, then, is a tactic that allows the institutional leaders of science to distance themselves from social and political issues by falling back on the methodological ethic which governs the production and validation of scientific knowledge, whilst abstaining from comment on its use. In other words, in the professionalized scientific community fact and value have become institutionally isolated. In keeping with the development launched by the Reformation, values are regarded as individual matters and only facts derived from valid methods can expect institutional support.

As the scientific outlook has diffused through society most professions have adopted an instrumental approach to their function and promoted technique as the key to advancing human welfare. For example, education and training in both the scientific and the science-oriented professions stress to the exclusion of almost anything else technique and methodology. Little attention is given in curricula to the social role of the professions or to the social problems with which they ought to be concerned.

The political and methodological dimensions of science are interconnected. We have seen that by promoting methodological purity individually and a tactic of prudential acquiescence professionally no social ethic has been developed. The danger, it is now clear, is that the scientific professions are extremely vulnerable to economic and political manipulation.

The Practical Dimension

This brings us to consider the final aspect of the contemporary scientific ideal: its practical orientation. Ever since Bacon first formulated his aims for the scientific project in *Novum Organon* science has been oriented towards utility. While there can be no doubt that today science values theory, still in the long run every theory is expected to throw light on practical issues and concrete situations.

Nowhere is this practical orientation of science more in evidence than in the research and development activities of industry and government. In virtually every country in the world, scores of scientists are employed by public and private corporations to work on problems as complex as developing a new missile system or as mundane as producing a less energy intensive toaster.

Now, I have referred to a practical dimension to science and that dimension refers not only to the employment of scientists as investigators of every conceivable situation but also, and more importantly, to the fact that science has now become a crucial instrument of national policy. As such, science itself has come to be regarded as a resource and, as with many other resources, is currently being subjected to rational analysis. In brief, the techniques of science are beginning to be applied to the analysis of science itself; and the aim of this analysis is seen to be to allocate a scarce resource as efficiently as possible.

This view of science as a resource lies behind the initial conception of science policy. In its earliest manifestations, science policy research aimed at little more than measuring the amounts of money that flowed into particular areas of scientific activity. Latterly, science policy analysts have become more ambitious and tried to develop national plans for science and to set out criteria for choice among the various fields of science given that the financial resources available are always finite. Although science policy is still in its infancy, enough research has been carried out to indicate that, in nearly every country, responsibility for funding science has passed largely to governments or large industrial organizations. As a result of this, science is becoming, like so much in the contemporary world, increasingly bureaucratized. Increasing numbers of the most able scientists are recruited to work in large organizations, whether public or private, where scientific priorities are determined less by free enquiry among scientists and more by the needs of the particular bureaucracy.

RECULER POUR MIEUX SAUTER

Let us now take stock of where we are. I have used the notions of an Aristotelian and a contemporary ideal of science to try to clarify a process of development. When Aristotle formulated his ideal in terms of 'true certain knowledge of causal necessity', he also provided in the Academy the institution for acquiring this knowledge. The atmosphere was to be one in which an individual was free not only from the cares of providing for his material well-being but also, and more importantly, free from the demands of political activity which was, of course, the highest calling in the *via activa*. In other words, it was understood that only by disengaging oneself completely from the *via activa* could one hope to encounter the deeper experiences of *via contemplativa* in which the true nature of reality was disclosed. It was by means of the Socratic method that one could be helped along the road to philosophy and, no doubt, it was in the Academy where the benefits of dialogue and, therefore, of a community of fellow human beings in pursuit of the truth was both discovered and sustained. Equally, there can be little doubt that this notion of an open community of scholars is no longer what is meant by freedom and autonomy in scientific research.

It seems to me that however useful it may have been for the early followers of Galileo and Newton to portray themselves as engaged in a philosophical enterprise analogous to, but more effective than, that of traditional philosophy the use of language of this kind has never been much more than rhetoric in the development of modern science. For the Greeks, science – because it contemplated the highest things – was static, individualist, necessary and theoretical. Contemporary science, as we have just seen, is dynamic, collective, provisional and practical. Therefore, it would probably be more appropriate, were it not for a certain linguistic inertia, to drop the notion of science altogether and to talk about research; and to denote researchers by their specialisms rather than talking generally about scientists. We have already seen that it is the job of specialists to keep their specialisms flourishing. To do this they need, at the very least, an organization which allows new blood to be continually recruited to the specialism and through which the resources can be made available to develop it. In Western Europe, this service has been supplied for the most part by institutions of higher education; universities, polytechnics, *grandes écoles* and *Technische Hochschulen*, etc. At first, of course, the costs of providing scientific training were slight because the numbers were small and equipment inexpensive. But, in line with industrial development, not only has research become more and more capital intensive but so has the cost of competing internationally. The result has been that the professional exchange relationship chosen by most scientists in the nineteenth century as a way of securing social support for their activities, has on more than one occasion become strained. Indeed, it is only when the relationship becomes strained as, for example, in the current recession, that it is possible to get a glimpse of the tension between the desire for each specialism to move according to its own autonomous dynamic and the need for it to provide trained manpower and a flow of knowledge useful to society. This more or less loosely coupled, confederation of specialists, we have argued, is more accurately described as a research system than an open community of scholars; the links are functional not existential.

CONCLUDING REFLECTIONS

Having suggested that 'research' is a more appropriate term than 'science' let us ask in what sense research is a commodity. Now, a commodity is 'any kind of thing that meets needs or is wanted or can be traded'. Science in the classical tradition, since it was an *individual achievement* involving not only a commitment to truth but openness towards the experience of Being, could not, in principle, be traded for anything else. It was valued for its own sake. But, the new science from the very beginning sought to undermine reliance on this kind of experience as a basis for authoritative knowledge. Descartes, with his injunction to distrust all things, spread suspicion into virtually every facet of knowledge whether of the senses, or reason or of revelation. But his success, as we know, rested heavily on the fact that he had discovered another model which, it seemed to him, was foolproof: mathematical reasoning. The important thing about this method, and this was recognized by experimentalists such as Galileo and Newton, was that it made it possible to 'make nature' according to our own ideas. Since, according to the dictum of Vico, the only things we can properly understand are those we make ourselves, the road was now open – though not all at once – to explore to what extent it was possible to reconstruct, via experiment, the processes which go on in nature. Indeed, it was not until much later – in the twentieth century when the technological skills were available to allow scientists to set up processes which do not occur naturally on the earth – that the term of this methodological development was reached. Thus, the philosopher of science Philip Frank had good reason to define contemporary science as 'the task of producing desired observable phenomena'.[10]

To the extent that research discovers processes or, with the help of technology, initiates them, researchers could be described as a species of *homo faber* – man, the fabricator. And it is as a fabricator that the researcher finds his closest affinity to the technologist and the industrialist who are also searching for processes, in their cases processes to stimulate or maintain production. This alliance is far closer than is suggested by the usual notion of technology as applied science, whereby the technologist or industrialist is supposed to apply discoveries made by science. There are, no doubt, instances where this occurs to the mutual advantage of both science and technology, but far more significant is the fact that both scientists and technologists use the same method: both are trying to make something; the former to gain understanding, the latter to gain some competitive advantage. But clearly, as the processes used by industry become more and more remote from experience, the more they are created technologically, the greater the importance of the researcher to the industrialist. So, the practice of science becomes like a commodity in at least two senses. In the first, and less interesting sense, research activity generates knowledge and provides units of skilled manpower which, because specialists exist in an exchange relationship with society, *must* be traded. This is the condition of contemporary social support. The second, and more important sense, lies in the nature of the exchange relationship itself. In a society which understands itself as a network of such relationships, each group derives its power and ultimately its value from the function it fulfils in relation to other social processes. And in an industrial society, as Marx observed long ago somewhat sadly, exchange value tends to prevail over use value. This, then, is the real significance of the commoditization of science: science – whether in the sense of the knowledge it generates or the skills it provides – ceases to be valued for its own sake. Instead, it derives its value from its utility in relation to production or some other social process.

If we link these ideas with what has been said previously about the contemporary scientific ideal, it can be seen that because of the exchange relationships which were created by the professionalization of science in the nineteenth century, the academic research system is now linked by a network of social obligations to government and industry. Insofar as government promotes material well-being via technical progress as a priority, the academic research system's role is to play its part in the process of industrialization and modernization. It could be argued, then, that by relinquishing the Aristotelian ideal and by adopting methods which would yield only hypothetical knowledge;

by establishing a form of social organization based on a professional exchange relationship; and by promoting an ethic that depends on methodological rigour to the exclusion of virtually everything else we have produced an academic research system more or less adequate to the needs of industrial society. The benefit, it appears, of adopting a method which generates only hypothetical knowledge is *purely* practical; it allows us to remake nature and produce the commodities we need to enhance our standard of living. The cost is the relativization of all knowledge and values. As Arendt has put it, since we are simply applying the products of our minds that act of production does not in itself carry any standard of evaluation.

We have reached a somewhat paradoxical conclusion. The development of the contemporary ideal of science leads to the notion of research as a commodity. Science, too, it appears, acquires its value in relation to the function it performs and, insofar as industrial values predominate, science is valued for the role it plays in production. Since Descartes urged us to doubt all things, there has been no other source of value. The paradox is that the commoditization of science should be seen as a *threat*, furthermore as a threat to the open community of scholars – a notion that was swept away by the founders of the scientific revolution in the sixteenth and seventeenth centuries.

In sum, then, the contemporary scientific ideal as currently articulated leaves science with no other value than its function with respect to social processes. Because of this, the only role for the academic research system lies in creating the knowledge and supplying the manpower necessary to keep society functioning smoothly and efficiently which today means via the continuous process of technological innovation. The threat to the open community of scholars appears to be a total anachronism. Are we not hearing, perhaps for the last time, the final vibrations of that 'fundamental chord' of our tradition referred to in the opening paragraphs of the paper? An open community of scholars still stirs our affectivity and we are annoyed – why? Perhaps because something of value is slipping away although we have no idea what it is or how to recover it. Could it be that in abandoning the Aristotelian ideal we may have thrown out the baby with the bath water?

If this analysis has some validity, then it follows that the juxtaposition of science and the open community of scholars refers to an element in the tradition that we have inherited from the Greeks and that that tradition has now been replaced by another which utilizes a different type of knowledge and requires a different form of social organization. In brief, the academic research system is the appropriate one for an industrial society. The role of the academic research system that has been developing during the last hundred years aims to provide useful knowledge and trained manpower for that society by applying, more or less successfully, methods first discovered in the sixteenth century. It is true that the theme of usefulness has been a recurrent one in this development and that it tends to recur particularly during times of war or economic depression. But the rhetoric that emerges on each of these occasions is not meant to *establish* usefulness as the role of the academic research system but to *develop* it further. When scientists decided to seek their legitimation from industry and their resources from the state it was on the premise of the usefulness of the knowledge. From the beginning, science has been organized to fulfil this task.

If there is, currently, a debate about how to fulfil this task it is about mechanisms and administration rather than about fundamental values. Nobody doubts that a social group which receives public support should offer something in return; the question is about how best to do that. The kind of freedom from the cares of material provision and, more importantly, from political activity in the Academy were meant to provide the occasion for intelligent men and women to recover for themselves the experiences on which Greek philosophy was based and so guarantee its survival. In our age, when we speak of freedom we refer to the autonomy of the specialism to pursue its own enquiries according to its own lights. And specialist autonomy can be guaranteed only in return for knowledge and trained manpower, ie, by its usefulness. Strange as it may seem, the element of randomness in scientific discovery is understood in most policy circles and so, therefore, is the need to fund science as a general social overhead on other more immediately profitable activities. The question of science as a social overhead is not challenged but what is debated, more or less continuously, is how large that overhead must be and how best it might be deployed. In other words, the debate is about how to allocate resources more efficiently; it is a difficult, but more or less familiar economic/administrative problem. So, it is wrong, in my view, to react as many have done during this current recession to the notion that government in its priorities is trying to make the universities utilitarian. The universities are already utilitarian institutions and, it is understandable that, in times of financial constraint, governments will try to develop policies to bring academic research institutions into line with available resources. And, because the more or less regular demands for utility from the higher education sector must be seen as part and parcel of the ebb and flow of political/

economic life, it is not accurate to regard this as indicating in any sense a changing role for the academic research system. What is going forward is mainly the development of a role whose function has been established.

To some of you, this may seem a depressing result. It will now be clear that because the notion of the contemporary ideal of science is a dynamic one, we can expect to see further specialization and professionalization to proceed apace; we can expect to see the distinction between the natural and human sciences disappear and be replaced by research concerned with mastery and control of the natural and human environments and a rump classified as 'others'. We can expect to see the academic research system continue to develop via new methods, techniques, models, new specialisms and, therefore, to devote its time in the university primarily to passing on skills which, having been found functional within the specialism, will be functional for social progress as well. In brief, we can expect to see the university as the 'reproductive organ of the *cultural* community' gradually to atrophy because culture is concerned with passing on traditional values and it is now traditional in contemporary science to oppose the traditional.

The changing role of the academic research system lies not in the fact that it may be becoming more utilitarian, but in the possibility, which is hardly yet a probability, of a deeper understanding of the limitations of specialism. At its root this involves making socially effective a method for thinking about praxis; about what use we want to make of our knowledge of nature and man. I say method because just as the legitimacy of science lies not in its results but in its methods and just as progress is possible because these methods can come into the possession of a group, so too the question of praxis is not about a bright idea to solve all moral problems but about developing methods adequate to these problems. Praxis acknowledges that we have come to the end of the age of innocence in which human authenticity could be taken for granted. Research in both the natural and social sciences makes it abundantly clear that our tradition contains a mixture of the rational and the irrational, of the authentic and the unauthentic, of the good and the bad. We are aware of all these elements when we experience, as we do now, the limitations of specialization; and this awareness has happened not only in the public-at-large but among the specialists themselves. If within each specialism there was an interest in the question 'what use are we to make of our specialist knowledge?', we could create within the academic research system the possibility of a continuous collaboration which could try to establish whether and to what extent its utilitarian stance was intellectually rational or irrational, whether morally it meets the requirements of authenticity and human development and in the last analysis whether the contemporary scientific ideal is, historically, an instance of progress or decline. Insofar as it becomes possible to identify what is rational, promote what is authentic in human development and search for what is good, research may acquire a value other than utility and to this degree science will cease to be a commodity.

REFERENCES

1. This quotation appears in the last of a series of four editorials by the editor of *The Times Higher Education Supplement*, Peter Scott, devoted to the 'Educational Revolution and Modern Society'.
2. Arendt H. 1961 *Between Past and Future*. The Viking Press, New York, p 18.
3. Some of these ideas are treated in Gibbons M 1982 The Contemporary Transformation of Science. *Manchester Literary and Philosophical Society* (new series) **1**:71. The notion of the classical ideal of science derives from Lonergan B. 1977 The Ongoing Genesis of Methods *Religious Studies* **6**:343.
4. See, for example, Storer W. 1966 *The Social System of Science*, Holt Rinehart and Winston, pp 77–80.
5. Haberer J. 1972 Politicalisation in Science *Science* **178**:721.
6. Bledstein B. J. 1976 *The Culture of Professionalism*. W. W. Norton and Company, New York, pp 80–128 and pp 287–332.
7. For some aspects of this development in Britain see McLeod, R. M. 1972 Resources of Science in Victorian England: the Endowment of Science Movement 1868–1900 *Science and Society 1600–1900* edited by P. Mathias. Cambridge University Press, pp 111–116.
8. Haberer *op cit*, p 722.
9. Haberer *op cit*, p 723.
10. Frank P. 1957 Philosophical Uses of Science *Bulletin of Atomic Scientists* **XIII** (4 April).

CHAPTER TWO

The Notion of an Open Scientific Community in Historical Perspective

Sheldon Rothblatt

In my discussion I would like to examine the two different sets of pressures that affect the way in which science as an organized and institutionalized form of intellectual activity is carried out. The first is rather dramatically identified in the title of this book. It consists of outside demands for a wide range of scientific services. Originating in government, in the military, in industry or more vaguely from the public, such demands are often considered inimical to either the best interests of scientists or to the nature of the scientific enterprise itself as it has evolved over the centuries. The assumption, fully contained within this book's title, is that science by definition requires absolute conditions of free enquiry in order to progress, to prosper, or to realize its true purposes and character, that scientists are the best judges of those conditions, continually seek them and are unified in their opposition to outside interference with the methods, problems, objectives and policies governing scientific research. Also implicit in the title is the assumption that if left to its own devices science would be pure. Thought is its own reward, the argument goes. High technology or applied science, the means by which science produces commodities, are necessary evils, the result of commercial, ideological and global conditions which scientists as a community are powerless to alter.

The second set of pressures is not alluded to in this book's title but in my opinion also requires discussion, for in a number of fundamental ways it casts light on the first. These are the pressures that derive from the internal constitution of science, from its cultural or value system and from the institutions that scientists themselves have built or have cooperated in building in order to maximize the conditions under which their work is performed. In more recent decades the words that incorporate the notion of an organized scientific community are 'profession' or 'professionalism', although the reality has existed for over the past hundred years. In describing the institutions of science, therefore, I will often be addressing problems that have arisen as a consequence of the development of science as a full-time career or paid occupation customarily classified under the service sector of the economy.

I will attempt to illustrate the ways in which science and scientists have responded to pressures imposed from the outside and how their own internal imperative has led them to take an active part in fashioning the institutional solutions which are at present much discussed. I will argue that scientists have never passively submitted to historical forces and circumstances on the grounds that they were somehow beyond control. In short, I will attempt to qualify without absolutely denying the assumptions that scientists *ipso facto* comprise an open community or instinctively resent pressures to transform science into a commodity. Any historical record is bound to be complex; but nothing is gained by oversimplifying the institutional and cultural arrangements that men and women undertake to bring meaning and regularity into their lives. If we can recognize some of the sources of confusion, we can also deal with them, at least to a certain point.

I will limit my remarks to aspects of the history of science in Britain and America. The treatment of the two societies is not even: it would be impossible to be comprehensive within the space allowed. Nor is it strictly comparative: despite a common inheritance and similarities in their histories, these English-speaking countries have differed profoundly from one another. There have been periods in which in some respects they resembled one another more than they do at present. By selecting examples and situations from particularly critical points in the history of the two nations, it is possible to draw special attention to the problems under discussion and the different historical contexts in which they occur.

I will only speak about free societies with established traditions of representative government, societies in which civil liberties are embedded in law, custom or doctrines of natural rights, where freedom of choice and movement exist and in which the right to criticize authority is either taken for granted or enjoys constitutional safeguards. The story would be different if I were considering totalitarian societies, where science is wholly an instrument of state policy and can be perverted, as we so well know from the history of Nazi Germany, into the most unspeakable ends. Free societies nevertheless undergo many phases and if we are to appreciate the development of the subject we are discussing, we should try to assess the impact of history on science as a form of cultural life freely entered into and voluntarily pursued.

THE PROBLEMATIC CHARACTER OF 'THREATS'

The distinguished scientist, I.I. Rabi, once called science a culture rather than a community. No doubt this is profoundly true. To speak about a scientific community in today's bewildering intellectual world is almost to speak allegorically. There is no single scientific community in either Britain or America, and to speak of an international community as if it had a centre or a core is to defy reality. If there are communities at all they are multifarious, representing different specialities, different scientific outlooks and different social and political viewpoints. And even within each of them splits and disagreements are evident. One simply cannot discuss 'threats' as if that ominous word in any way suggests tight organization and general agreement. Whatever constitutes a threat to one group may be interpreted quite differently by another. That is certainly true now.

It is also true of the past. It is particularly true because for several centuries the occupational standing of scientists was nebulous, and it is awkward for us even today with so much hindsight to determine the precise relationship of one group of scientists to another. The word 'scientist' itself, as is so often remarked in the history of science, is a British neologism of the 1830s and 1840s, quite possibly coined by more than one person but certainly attributable to the eminent Cambridge University philosopher, William Whewell. It was by no means eagerly seized upon. The word dragged on for decades unappreciated, the butt, in fact, of much humour and invective. It was denounced for being a formation from incorrect Latin or a Latin –Greek hybrid. British writers particulary enjoyed calling 'scientist' an Americanism (as they called anything they disliked an Americanism), as unpleasant a word, said the great Professor Thomas Huxley, as 'electrocution', which was, in fact, of American coinage. 'Scientist' was not accepted in either country until the nineteenth century was nearly over.[1]

In the eighteenth century individuals interested in science were not necessarily careerist, as I will explain later. They went by many names. They were practitioners or natural philosophers or academicians belonging to academies. They were cultivators of science. Even the word 'science' was undifferentiated before the nineteenth century and was interchangeable with 'philosophy'.

In France there was similar confusion. *Le scientiste* applied only to the followers of the nineteenth-century school of scientific materialism. *Un scientifique* was not commonly used until the twentieth century. The favourite old regime workhorse word was *savant*, which pulled in meanings from *érudit*, but it had no specific occupational reference. Those who wrote scientific texts in France (and in England) were known by the catchall designation of man of letters.[2]

The commitment of eighteenth-century individuals to science varied greatly. To many of them science was a hobby or amusement. The century was filled with charlatans, *jongleurs*, millenarians and assorted enthusiasts who hid behind the mask of natural philosophy and enthralled audiences with sensational electrical spectacles that came close to reviving superstitions the Enlightenment had been eager to dispel.[3] John Heilbron has suggested to me that this motley group might be called purveyors of science. Another group, composed of members of the learned societies or holding professorships, experimentalists with institutional affiliations and a demonstrated interest in advancing science, can be called doers of science. For some parts of my discussion I will employ these designations, crude as they perforce must be, provided at the same time it is undertstood that there was crossover between the two groups. They belonged to many 'reference groups' as we might now say, following the lead of Robert Merton, and could hardly develop a strong community consciousness. Therefore, we should not expect at that early date to uncover the linked characteristics or fundamental career interests denoted by the words 'professional community'. The fluid societal personality of the scientist must be constantly born in mind as we attempt to understand the effects of historical change on the practice of science and on science as an organized social activity.

SCIENCE AS A COMMODITY: THE IDEA OF THE MARKET

The phrase 'science as a commodity' suggests a consideration of what we commonly refer to as market forces as they influence the structure, organization and context of scientific activity. It leads us to analyse science as more than a system of thought and reasoning akin to other systems of thinking, eg, history, economics, linguistics. When we refer to science as a commodity we are being asked to regard it as a product of exchange and detachable from the mind that gave it life. When some years ago I first heard intellectual work described in this way I was quite surprised. What began as dismay turned to contempt and then gave way to reluctant sobriety. I had always believed the work of the mind to be unpurchasable. But of course that was naivety; for the purchase of men of thought has a very long history, as long as any. Men of ideas have often sought prominent roles in public life, and have turned from the world, as did Epicurus, when their ideas were incapable of ready communication. Pythagoras debated the problem of philosophy, that is, whether to participate actively in the politics of his city-state, Croton, and he turned towards seclusion and the pursuit of pure truth, as much perhaps from frustration as from the attraction of ideas in themselves. Thought is no more immune from outside influences and pressures than any other product. Nor is it autoimmune, secure from the ambivalences and contradictions inherent in its own value and delivery systems. Foremost among those ambivalences is discomfort with the idea of market operations as a factor in the generation of knowledge and discovery. Even some economists of higher education are troubled by the crass implications of supply and demand and prefer to talk about intellectual work as an investment rather than a consumable product to be discarded and altered as fashion requires. As an investment the return is deferred. In the interim, the investigator works at his own pace, or at a pace determined by his peers. He must still meet deadlines, he must in some way still be accountable for the support he receives; but he is also given some protection from short-term fluctuations in the market.

What effect does the market phenomenon have upon science? A proper answer would have two parts. The first would indicate the effect of market forces on science as a system of reasoning and understanding, on scientific method and the rules governing evidence. This is elusive and hard to measure. The deutero-platonists might argue that the market can have no effect on the processes of thought, but that is debatable. The second part of the question yields a clearer answer. Whatever the influence of the market on cognition, it dramatically affects the institutional structure of science, the types of research undertaken and the questions asked, the purposes for which research is assumed, the system of incentives essential to its continuance, the motives for embarking on scientific work in the first place, its quality, the value placed upon it, and the physical setting in which it is carried out. The market does something else equally dramatic: it

25

transfers or threatens to transfer decision-making and evaluation from one kind of authority to another, so that those who produce and those who consume may have opposed value systems although brought into some temporary form of accommodation through negotiation and mediation.

What is the relationship between the market and the open community of scientists? It is possible to argue that in theory they are in opposition. The idea of an open community does not suggest competition for goods and services, as does the market model, but the very opposite: insulation from demand, cooperation among peers, the free exchange of ideas, science as a shared system of values. It suggests that thought obeys or should be allowed to obey its own rules, the 'rules of the mind' as Victorian moral philosophers would have termed it, and the rules of the guild or the academy. In its pristine form the market principle embodies the notion of individualism, which partly springs from the atomic materialism of the Enlightenment. It suggests 'natural' competition for recognition and reward. The notion of a community of scholars or scientists implies quite the contrary, a universe in which the whole is greater than the parts and in which knowledge is the supreme attainment. In this ideal boundaries and barriers do not exist. National, group or personal self-interest is deplored.

The claim that knowledge has intrinsic worth has the authority of Pythagoras and Plato, and the choice between collaboration or community and competition likewise has a history. Both have been used during periods of acute social rivalry or wherever traditional values seemed threatened by new ones. Both acquired considerable rhetorical and polemical force over the centuries. Scientists repeated the arguments, as did scholars, writers and social scientists, and they appear in other contributions to this volume. They have played a part in the self-definition of one kind of modern intellectual. There is always some truth in polarities, but the advocacy of an ideal also requires that worst case scenarios be argued so that the issues are clear. There remains a difference between ideal and actual states, and if we are to evaluate the meaning of the market for the doing of science, we have to take historical circumstances into consideration.

HISTORICAL DEVELOPMENT OF THE MARKET: ITS EFFECT ON INTELLECTUAL LIFE IN BRITAIN

The theory of the market was developed in the late seventeenth and early eighteenth century and is closely associated with the peculiar economic environment. Britain first became an international trading power in the late seventeenth century after a series of commercial wars with the Dutch. Because of her climate, a shortage of vital natural resources, and a geographical location as an island nation off the coast of Europe, she was never autarkic as, say, France, and therefore the mercantilist or protectionist philosophy of exhausting rivals by small wars never really worked as well in Britain as it did elsewhere. From an early date, virtually without being conscious of it, Britain adopted a *de facto* free trade economy encased in a thin shell of protection. She lurched towards a balance of payments economy and away from a bullionist monetary economy. The result was the creation of a vigorous trading and financial community located in the City of London, a representative selection of foreign trading interests nearby in another district, and a habit of acquisition frankly oriented towards the open market. This development was abetted by an influential section of landed aristocrats who, for quite independent reasons, had taken control of ministerial government. They needed the 'monied interest' (as it was known) in order to solve certain revenue problems of a structural nature. Thus commenced a collaboration of propertied and trading interests absolutely essential to the constitutional history of eighteenth-century Britain. As national income grew, so did other linked characteristics of a market economy. The British aristocracy, grown wealthy on the profits of land, office and City investments, became heavy consumers – aristocracies are oriented towards conspicuous consumption anyhow.

The increase in aristocratic wealth and power reinforced the system of patronage that had long been a feature of British society. Throughout Europe, in fact, the sponsorship of intellectuals by noble and regal courts (and by princes of the Church) was a long-standing responsibility of privilege. Of course patronage can be regarded as a form of market economics. Under conditions of patronage demand is relatively inelastic, the market is restricted, and the supplier has limited opportunities for selling his product. When we speak of the market, therefore, we mean primarily the open or unrestricted market with unlimited competition and a more dynamic relationship between supply and demand than obtains under conditions of royal monopoly and noble patronage, where the seller is usually at a disadvantage. We mean also that the relationship between buyer and seller is impersonal, regulated, in a famous phrase, by an 'invisible hand'. No other consideration governs the connection except profit. Friendship, deference, coercion, personal obligation, religious moral scruples are absent in the theory of the open market although never wholly absent in real life.

In the early eighteenth century the open market began to challenge patronage as a means of providing a livelihood for writers, artists and thinkers. It was the great economist, Adam Smith, who stated the significance of the market for intellectual work in unmistakable terms. 'In opulent or commercial societies,' he wrote, 'to think or reason comes to be, like every other employment, a particular business which is carried on by a very few people who furnish the public with all the thought and reason possessed by the vast multitude that labour.'[4]

Of the two competing systems of sponsoring intellectual work in the eighteenth century, initiatives connected to the market produced the more radical departures in art and thought. It is perhaps only necessary to list them: the social novel, a form of literature denounced at the time as 'romantick', that is, frivolous and insubstantial; the so-called British School of Painting, featuring portraiture, since buyers wished to be painted; English opera, a popular form of opera as distinguished from Italian opera which was the importation of foreign models; connoisseurship, that is, the collecting of objects of all kinds for pleasure and as a hobby and the production of writings pertaining to the art of collecting. The hobby, in fact, is very eighteenth century. Gardening, as one example, created an immense industry of growers, nurseries, landscapers, builders of garden ornaments and tools, collectors of seeds and plants from the world's exotic zones. Music also went into the market and was composed especially for outdoor pleasure gardens or Tivolis.

As consumables, the arts underwent a process of packaging or marketing. Inevitably, given the existing variety of markets and publics, a certain amount of popularization and vulgarization took place. Polemical or controversial writing, a notable result of aristocratic in-fighting early in the century, declined in importance. There was a conspicuous increase in ephemeral writing and a marked demand for entertainment and recreation.[5] Music was supposed to be pleasing, novels diverting, amusing or mildly horrifying (the 'sublime'); gardens were for recreation, for strolling or riding, hence the invention of the English park of trees, shrubs, lawns, bridle paths and lakes. In these circumstances some writers and painters preferred the patron to the impersonality and vagaries of the market. Others took the paths of least resistance and accepted the market. Still others, accepting art as a commodity, learned to manipulate the market. At least they attempted to elevate public taste while at the same time making moral and aesthetic lessons palatable. The novel became didactic, even while it had to be humourous. Painting became neoclassical, by adopting the technique Santayana has called the 'idealization of the familiar', wherein heroic, epic, religious and historical models were used to enlarge the perspectives of sitters. Even the foolish gothic novel of the late eighteenth century – usually the story of a late adolescent woman undergoing a series of improbable adventures – was an attempt to heighten sensibility and even to introduce an element of religious feeling into works meant essentially for amusement. For artists of greater talent the market was a constraint but also an opportunity to innovate and to experiment; and out of their attempt to please, yet have matters their own way, came some distinct literary achievements, some splendid architecture, some excellent painting and a new conception of the aesthetic.

There were restricted as well as open markets for science in the eighteenth century. There were markets for certain kinds of practical talent. Applied mathematicians found jobs in government, as did surveyors and cartographers. The doers of science preferred restricted markets as less volatile and providing some security of employment. Their homes were institutions like the Royal Society and teaching academies established by non-Anglican Protestants, that is, upper secondary schools in current European idiom. The universities of England and Scotland provided some opportunities for advancing science. But the doers of science are occasionally also to be found in the open market with the purveyors of science, sometimes driven there by necessity. The purveyors were most in evidence however. They were popular lecturers, instrument makers, writers of best-selling texts, busily creating a mass market for scientific and technical information and for scientific products. Compilers of almanacs did a thriving business. Provincial literary and philosophical societies disseminated scientific ideas and gadgets to town gentlemen. Other forms of intellectual work show the effect of the spread of science as a model for thought and reasoning, for example, Scottish social science. The greatest animal painter of the age, George Stubbs, was something of an anatomist.

Science was popular in the eighteenth century, but it also underwent the same sort of trivialization that affected other kinds of intellectual activity. Astronomy at times appeared to be a party game for fashionable audiences enchanted by views through the telescope.

Sensational experiments with Leyden jars and air pumps were conducted before titillated audiences. Fashionable gatherings in France
were captivated and delighted by mesmerism. Writers on chemistry
sometimes justified its study on the grounds that it was a polite
amusement, suitable for light-hearted conversation. In general the
market did not have quite the same revolutionary effect on the structure of support for science as it did on the literary and plastic arts,
where supply and demand factors eventually eliminated patronage as a
serious means of sponsoring intellectual talent and led to the rise of
new mediators between the artist and the public, such as publishers
and gallery owners. Compared to European science in the seventeenth
and eighteenth centuries, British science had a rather meagre support
base. The Church of England, for example, provided no direct institutional support comparable to the assistance given Jesuits by the
Society of Jesus, which was the leading patron of physical and
mathematical science in the seventeenth century. Nor did British
societies or academies dedicated to the advancement, improvement or
cultivation of science like the Royal Society provide as many paid
career positions in the eighteenth century as did those in Protestant
Germany. A certain amount of significant intellectual work in experimental physics was carried out by professors in Oxford and
Cambridge and in the Scottish universities, but their role was
essentially the traditional one of teaching or professing rather than
research or free enquiry. The commitment of British professors as a
body to the ideal of discovery was precarious. For those wishing to
contribute to the progress of science, which meant finding extra
income for the purchase of equipment and assistance, there was the
expedient of private lectures arranged on top of all other official
duties. The universities in some European countries, the Netherlands
for example, managed to incorporate a research mission into their
institutions. Interestingly enough, because the Dutch universities
were responsive to new trends, the introduction of scientific
academies was delayed.[6]

All but a handful of the most itinerant 'electricians' found the
market for lectures inadequate for their financial needs and sought
other ways of adding to income. Occasional teaching in secondary
schools, surveying, instrument-making and the writing of textbooks
are mentioned in eighteenth-century accounts. Tiberio Cavallo (born
in Naples) and William Nicholson were particularly successful writers
of popular scientific texts.[7]

A conspicuous feature of the support structure of eighteenth-
century British science is the substantial number of practitioners

whose primary source of income was not from basic or pure science, was not a salary or stipend derived from a scientific institution or position, but came as the result of following another occupation. Professional men such as clergymen and physicists were interested in science even if there were few opportunities for them to make a living entirely from this source. Government agronomists, astronomers and cartographers were also able to pursue scientific interests, although not with absolute freedom. Those who were freest to indulge their scientific curiosity were rentiers and other men of independent means who took no heed of market demand since their income was assured. The presence of such men, as well as the active participation in scientific activity of those who did not make a living from it, has led an older generation of historians of science to lump them all together as amateurs. More recent historians object to the word. Instead they categorize practitioners according to their scientific contributions and the degree of respect accorded them by contemporaries. But all historians essentially agree that the professionalization of science, its development as a distinct career specialty, requiring new forms of institutional differentiation, new standards of qualification and a different support structure, is more properly a nineteenth- than an eighteenth-century story, some regard being paid to antecedents and survivals.

THE MARKET FOR APPLIED SCIENCE AND TECHNOLOGY

Thus far I have avoided direct discussion of technology and applied science. These are the forms in which most people understand science and for good reason. Technology and applied science are visible. They are products, and they are consumable. They lie at the centre of the controversy over the allocation of resources to science. Scientists engaged in fundamental research often complain that the demand for new technologies inhibits the progress of pure science and distorts the overall science effort. The distinction between pure and applied science, however, may well be overstated. I will return to this problem later.

Historians have not agreed on the influence of basic science on technology, although they are generally of the opinion that until the eighteenth century pure science learned more from technology than it returned. Through the centuries technology was the result of craft traditions and trial and error. It is possible to trace some eighteenth-century inventions to theoretical work, however. The lightning rod was the result of the investigations of eighteenth-century experimentalists. The improvement of windmills and the design of ships' hulls owed a great deal to the calculations of mathematicians.[8] There are also indeterminate areas, such as the invention of James Watts' steam engine, the premier machine of the industrial revolution. Some authorities attribute it to Joseph Black's theory of latent heat, but others dispute this. D.S.L. Cardwell, who is always interesting on the institutional settings of science, stresses the connection between Watt and Black – Watt having been Black's laboratory assistant – but he also notes that the general state of many scientific fields in the eighteenth century was such as to rule out possible industrial applications: the phlogiston theory in chemistry, for example, would have caused more problems than it could have solved.

Several scholars have recently argued that a lively interest in applying theory to the industrial arts was emerging in Scotland, the northern English provinces and southern Ireland at the end of the eighteenth century. The members of the very short-lived Academy of Physics at Edinburgh carried out research on soda, and University of Glasgow professors studied bleaching and other industrial processes. The evidence does not suggest great utility as a result of these re-searches, however, many of which were classificatory.[9]

The Royal Institution, a new turn-of-the-century science association that gave a chair to the distinguished Humphry Davy, supported work in agriculture and tanning. The Royal Institution is an interesting example because it is one of the very few instances of an institutionalized response to an actual demand for applied science. Its distant origins lay in an earlier philanthropic organization founded to pacify the rural poor at a time of mounting disorder in the English countryside. It was the specific creation of a group of aristocratic landlords who were anxious to improve crop yields and animal breeding. Although Davy's prolific investigations, which he released publicly as lectures and subsequently published, did not actually lead to any practical changes, his benefactors were apparently pleased by his efforts to market science. Davy called himself a compaigner for science. He often treated science as a commodity, and was quick to employ utilitarian reasons on behalf of science when he needed money. His patrons allowed him to build a large voltaic pile, for example, when he contended beforehand that soil fertility directly benefited from electricity.[10]

These efforts did not go very far in connecting pure science with industrial development; but they do suggest a respectable start and are perhaps important, as one author maintains, in detaching portions of science from a late seventeenth-century equation with natural philosophy.[11]

DRAWBACKS TO THE MARKET

Some doers of science and all purveyors of science probed the market where survival, reputation, status or career was an issue. The recognition of open market forces in the eighteenth century converted mental work into an item of exchange and subjected many writers, artists and purveyors of science to the laws of supply and demand. The open market was not the only kind of support for intellectual work, but it was definitely believed to be an alternative to other forms of sponsorship, such as institutional affiliation and patronage. Whether it was thought superior is a nice question. That celebrated man of letters, Dr Samuel Johnson, thought it had a distinct advantage over patronage, which required deference and fawning, but he was reasonably successful. Others, leading a precarious existence, preferred a more reliable source of income.

Acceptance of the market produced mixed results. There was considerable diversification of mental and artistic endeavour. There was also work that was unimaginative, shoddy, cynical and frivolous. Market pressures resulted in plagiarism (the word had great currency in the eighteenth century) and other forms of unacknowledged borrowing. Competition led to conduct that was morally suspicious, actions that were regarded as reprehensible, sometimes because they were unprecedented. From the standpoint of contemporaries, there were losses and gains. There were casualties, and there were also individuals who learned how to manipulate the economy through advertising, prestige selling, the mass production of cheap copies of expensive originals and the shrewd use of middlemen.

That the development of a market for intellectual goods should have such mixed consquences is simply a reminder of its complexity and protean characteristics. Changes in taste led to variations in demand over time. More than one market could exist at any given moment. Thus the market for lectures on experimental physics was better in London in the early decades of the eighteenth century than some thirty or forty years later and better in the English provinces in the second half of the eighteenth century than in the London of the same period, despite high admission charges.[12]

For the independent researcher – and this applies equally well to the scholar as to the scientist – the greatest drawbacks of the market are instability and uneven demand. The process of creativity requires patient benefactors. The researcher wants discovery to proceed at its own pace. The argument often advanced is that there is a certain mystery to discovery, an element of unpredictability about it. Changes of emphasis or direction are not easily foreseen. Crises and paradigm shifts occur almost unannounced. Advances often result from the use of new and untried instruments. Even when a group of scientific practitioners embraces a teleological view of the universe, as in the case of the knot of early Victorian mineralogists and geologists at Cambridge University, everyday enquiry continues as if no specific end were anticipated. Concentration of resources and mobilization of scientific manpower may in particular examples hasten the process of discovery, crash programmes may accelerate the application of science to specific social and technical questions, but even here the risks are many. The expected solution may be elusive, the state of the economy may be disturbed, the total scientific effort of a nation may become distorted, the balance between basic and applied research thrown out of kilter.

The last references are of course to the 'big science' of the twentieth century. The mobilization of science resources on a scale such as the Manhattan Project or the moon shot were only utopian dreams in centuries past, and the savage wit of an anti-Enlightenment writer such as Jonathan Swift was sufficient to expose their absurdities. In *Gulliver's Travels* Swift travestied the Academy of Lagado (actually the Royal Society of London) and its researchers, the professor who had been 'Eight Years upon a Project for extracting Sun-Beams out of Cucumbers, which were to be put into Vials hermetically sealed, and let out to warm the Air in inclement Summers,' and another in-vestigator, described with Swift's shocking scatological humour, who sought to break down 'human Excrement to its original Food, by separating the several Parts, removing the Tincture which it receives from the Gall, making the Odour exhale, and scumming off the Saliva.'[13] Typically, Swift's method of combat is to equate folly with filth.

The independent British scientific practitioner of the eighteenth century, if without private means, did not have an adequate support base and certainly not by existing European standards. For him the market was disappointing. Aristocratic patronage was equally unreliable, as was manufacturing, since no continuous, long-term linkages between science and applied science had yet been made. The obvious solution lay in broadening, extending and adapting the existing educational bases of support. The university commitment to research could be strengthened if the professorial role were altered. New scientific organizations could be established, new universities, medical schools and technical colleges established, and the contribution of government to the maintenance of science systematically improved. In America at a later point in the nineteenth century the private foundation would prove to be a valuable sponsor. This long-term 'historical strategy' was not exactly a repudiation of the idea of competition for resources under market conditions, for scientists would still have to market their products, although in receptive economic environments. Institutions and governments could still compete for scientific talent and for the products of scientific effort. After all, they had done so since the Renaissance. Outstanding Enlightenment experimentalists like Christian von Wolff were in great demand throughout Europe and offered their services to the highest bidders. Their scarcity value was recognized.[14] Under such circumstances demand outdistanced supply, the ideal market situation from the seller's point of view.

LONG-TERM RESISTANCE TO THE MARKET: THE CASE OF HIGHER EDUCATION

The market as an idea and as a phenomenon was welcomed by many Enlightenment elites as an alternative to the regulation of intellectual life by Church or state, and in the nineteenth century market responses were built into the foundation of certain new universities in Britain. The great expansion in higher education that took place in nineteenth-century Britain was initially a private sector phenomenon. London, Durham, redbrick and later the London School of Economics were all private ventures, deliberate efforts to avoid the relatively limited response of Oxford and Cambridge to new publics. But as private ventures these failed, or were believed to be failing, by the time the nineteenth century ended. The expansion of the higher education sector in nineteenth-century Britain simply outraced demand. (It often did in America as well. The number of failed educational institutions in the United States in the past and even today always surprises foreign observers; but acceptance of the market requires failure.) When this occurred, anti-market sentiments surfaced. These had been present all along but had lain fallow. They were the particular result of centuries of royal charters and aristocratic patronage and government. Oxford and Cambridge had all along resisted the open market, although never with quite the indifference suggested by some writers. Victorian academics understandably began to prefer the centre to the periphery, to use Edward Shils' handy metaphor. Rather than undergo the potentially disastrous cyclical movements of the market, which interfered with steady growth and undermined stable support for expanding disciplines, they gradually transferred their attention to the state. Some scientists and scholars expressed reservations, fearing the consequences of too great a dependence upon a single source of research support, but in the uncertain educational environment of the late nineteenth century the state appeared more relaible than the market. It was in this period as well that the sentiment took hold that pure or disinterested scientific work was only possible where research was regarded as a form of sacred knowledge, to be nourished and protected against vulgar demands for instant results and practical discoveries. Researchers assumed that industry or public opinion was not sympathetic to theoretical work; and although, as we shall learn from a later discussion, interesting and fruitful ties between scientists in universities and industrialists did exist, their association was never successfully consolidated nor was it to become dependable.

The centre outlook of High Victorian culture has carried over into the twentieth century, and this has been agreeable to the leaders of both political parties, who tend to be dirigist. Even the thirty existing polytechnics, combining the functions of California community colleges and American four-year state-supported, non-research institutions, have sought the relatively protected market position of the university sector. Financial control over these institutions is a constant tug of war between local authorities, which respond to changes in political opinion, and the Department of Education and Science, which is directed by Whitehall mandarins but is not of course totally immune from the political oscillations of Westminster.

In America a federal constitution, a society of relatively open mobility and deeply ingrained anti-European sentiments produced in the nineteenth and twentieth century a culture more receptive to the notion of thought as an item of consumption. But these contrasts are hardly satisfactory, for in America too in the course of the nineteenth century a learned class began to experience the same sense of revulsion from the market characteristic of Europeans. They were afraid that changes in consumer taste would reduce the influence of the traditional curriculum or result in lower standards of achievement. The solution that eventually emerged – one that is in perfect keeping with the pluralism in American culture – was a highly differentiated and diversified educational base with room for all disciplines and opinions, for practical education as well as the most esoteric theoretical enquiries, for pure as well as for applied science. Educational institutions responded to national, state and local markets. Some colleges and universities were relatively independent from fluctuations in the market, others were totally dependent, and still others contained a mixture of responses within the same institution. Generally speaking, it was the elite institutions that were the most insulated from market controls. New institutions, small private colleges with weak endowments or public universities dependent for the most part on revenues derived from state taxes proved to be quite adaptable, adjusting curriculum and the recruitment of faculty to market forces. Regarded by elite institutions as occupying the periphery, market-oriented colleges and universities responded by praising their flexibility and by welcoming the open market for the opportunities it provided for the exercise of imagination and initiative. They accepted challenges in a Robinson Crusoe mood of exhilaration at the prospect of new territories waiting to be mapped. And they looked upon the centre as pampered, stodgy and timid. They accepted forthright the basic premises of the European and American Enlightenments that a dynamic society required a correspondingly dynamic educational system. If the established institutions behaved defensively with respect to changing patterns of demand, then the periphery had to assume leadership. Such was the reasoning.[15]

Resistance to the open market leads to the search for other sources of support. This is easier to achieve where traditions of state dependence already exist and where a learned class has the necessary network of associations and contacts to design its own rescue.

THE SEARCH FOR PATRONS

The diversification of the science support base in the nineteenth century is one of the really extraordinary aspects of the sociology of science in the past century, especially when contrasted with what went before. I fear the story is too complex for adequate treatment in the body of a paper such as this, but at least I can indicate the main directions of the change, the reasons for it, and the ensuing consequences. What we must give up – should we have the slightest desire to retain it – is any notion of science and scientists as passive, as busily pursuing their work with no other consideration but the intrinsic pleasure of the intellectual task itself. That there is an intrinsic pleasure I will not deny, but that the pursuit of knowledge can be discussed apart from all the other pressures that affect human activity even of the most cerebral sort no one, I am certain, would defend. In this the economists have a point. Intellectual work must be supported by an economic surplus. How that surplus is sought and allocated is of great significance in discussing the autonomy of intellectual work.

The fact is that nineteenth-century scientists of every kind were energetically engaged in seeking what were called 'endowments of research' and in broadening their institutional support base. The reasons for doing so were manifold. Besides the exigencies of the market, there was the increasing expense of doing science, of equipping labs, hiring assistants, maintaining buildings and outfitting expeditions. These costs were of course relative. In the mathematical fields the work was essentially abstract. Even in chemistry, metallurgy, and agriculture the costs were minuscule compared to what is required today. To properly explicate this theme the historian would have to take up each of the sciences, follow their trans-mogrifications and internal history and graph the moving line of costs. Even this would be insufficient. He would also have to identify the stages by which each of the sciences became professional, that is, became occupations which engaged the full energies of the scientific practitioner. By no means was this achieved uniformly throughout the sciences or throughout any of the areas of scholarly work. The historian would have to consider professionalization in relation to the question of social mobility. Were scientists increasingly recruited from families of lesser income, as in the case of late eighteenth-century writers? I am not aware that the subject has been examined extensively. There is some work to indicate that this was the case with distinguished mathematical graduates of Cambridge University from about 1830–60, but the evidence is not decisive enough to spin out the argument; and furthermore, most of the honours students or 'wranglers' did not enter science careers.[16] Oxbridge backchat from the late nineteenth century had it that science students tended to be drawn from families lower in the social scale; but this is precisely the kind of snobbery we would expect in a period when science was pressing upon classics for institutional support. There is no hard evidence to support what was in fact a jibe.

Assuredly one decisive factor was the necesssity of institutionalizing science teaching. Once an activity becomes professional and discards its craft or amateur skin, it automatically looks to education as a means of building up the field, choosing successors, and rendering its own status more permanent. This was happening continually throughout the nineteenth century. The medical professions led the way. City conditions created vast epidemiological problems which both engineers and physicians attempted to alleviate, often in competition with one another. Increasing per capita income and the growth of a very large urban middle-income sector led to a widespread concern for health and healthful recreations and created a bonanza for physicians after about 1850. When medicine returned to the English universities, from which it had been virtually absent for well over a century (the Scottish universities, on the other hand, could claim a flourishing medical school at Edinburgh), it became the umbrella under which new scientific subjects entered the university eg, physiology, bacteriology, medical physics and organic chemistry. Medical men were the leaders in finding institutional support for science. They were the dominant group in a scheme to establish a Royal College of Chemistry in 1845. They founded medical schools in provincial cities, threw their support behind the London University movement, improved the clinical capacities of existing hospitals, and in general deserve to be recognized as the most moving force in the institutionalization of science in the first half of the nineteenth century.[17]

Physicians and allied practitioners exploited and broadened a growing market for medical services. They used the licensing power of the state to eliminate the quacks and bogus medical men who had flourished in an unregulated market. They established a permanent home for themselves in both new and existing institutions and built up the numerous fields which today make up the medical specialities.

I will not enter into the popular pastime of debating whether medicine is a science or an art. Obviously it includes all the definitions from pure science to technology. Because of this versatility its uses were readily apparent to consumers, although confidence in the profession had to be publicly established first. This was partly achieved through the efforts of medical men to raise standards and improve qualifications, partly through their ability to regulate entry into the market and partly through key discoveries such as the germ theory of disease. Other scientific subjects were not so conveniently situated with respect to the market, yet in the British Victorian period there was quite a robust effort on the part of many different types of scientist to encourage outside support by offering services in return. The activities of professors in the new universities in the great provincial cities and the metropolis are notable in this regard.

The universities founded after the middle of the nineteenth century have diverse origins, but one common characteristic was the efforts made by the heads of departments to establish consulting and research links with local industries. This was made possible in the first instance by the spread of the research mission, which was incorporated into the universities only towards the end of the nineteenth century. The carrying out of that mission, however, required the cooperation of local industries, and this in turn awaited important changes in the structure of enterprise which occurred in the later nineteenth century. The older textile industries which had given Britain her industrial supremacy were small-capital ventures with limited technical requirements. The new industries of the imperial period were rubber and petroleum, the internal combustion engine, the pneumatic tyre, electro-communications, synthetic dyes, and electric power and traction. These required some greater degree of correspondence between pure and applied research. Furthermore, the new large-scale industries based on chemical technology required more technically trained executives, which the older family-run and limited partnership firms did not. Michael Sanderson's extraordinary book on the relations between the university and business sectors in the last hundred years provides us with a detailed and fascinating account of the role played by university-based scientists in British industrial development, and it is to a great extent a corrective to the widely believed argument that the contacts were minimal and insignificant. In later works he has also gone on to show that University of London professors carried out secret research for the British steel industry, and he has suggested there may have been more such agreements.[18] Clearly there is work to be done.

Any summary of the applied research efforts of the redbrick and London professoriate, however, does not go far enough in explicating our theme of science as a commodity. What we have to consider are the pressures that brought science and industry together in the late nineteenth century. This cannot be answered without an examination of the careers of investigators and the records of the firms with which they worked, and obviously I cannot do that here. I can try to describe the environmental and cultural conditions under which the scientists worked, however, to see if their decisions were affected by particular constraints.

We must note that the new universities of England were under-funded from the start. Start-up costs were exceedingly high, and expenses soon outraced endowment income. Operating expenses depended, in fact, upon fees. But the more modest income levels of civic university students made the revenue base uncertain and inadequate, especially as so many chairs had not been provided with endowments suitable for the style of life of a Victorian professor. Here, as in so many other instances in the history of higher education, the supply of educational services out-distanced the demand for them.

The private secondary education sector in Britain had ties to Oxford and Cambridge, but the expansion of the civic universities actually preceded the spread of state-supported secondary education, especially upper or pre-university courses. Their catchment areas were not national but local. Logically, perhaps, the expansion of higher education should have awaited the establishment of a thriving preparatory sector, but providing more opportunity for further education was by no means the sole reason for the creation of the new universities. Just as relevant was their symbolic and prestige value in the eyes of municipal leaders. The new universities were a part of the coming of age cycle of the great but undeniably recent manufacturing centres. Municipal leaders believed that a great city must have a university, just as it must have a city hall in the neoclassical and beaux arts styles. Something of the same view existed elsewhere in the world. German princes and the petty baroque autocracies of the eighteenth century founded universities partly for the prestige attached to having them, and civic pride required that the same effort be made by every American state. Status factors were probably as significant in the founding of new English universities as the desire to educate a reasonable proportion of the relevant age cohort.

Those who did come to the new universities were far too often ill-prepared for university-level work, and they were also of high school age – a situation which had existed in the Scottish universities throughout the first half of the nineteenth century. Faced with these structural difficulties, the provincial professoriate generally found that it could not carry on institutional research and advanced teaching where rudimentary science was demanded.

The professors turned to industry. Elsewhere, in laboratories such as the Cavendish at Cambridge and the facilities at Manchester, the professoriate found a different kind of market to support their research effort. They were dependent upon the fees charged for teaching engineering and medical students. Teaching was undoubtedly time-consuming, but it provided the laboratories with greater independence from the business sector and consequently a certain freedom to pursue more purely theoretical work.[19]

In the period before the Second World War there were signs that the alliances struck between redbrick and British industry were beginning to break down. The new leading sectors of British industry were very large and in a position to consider establishing their own research laboratories, thus reverting to the older British attitude that industry is better served by scientists directly employed in the works than by outside specialists possessing a greater degree of independence and perhaps freer to challenge priorities set by the firm.

At this point the American case is not, on the surface, profoundly dissimilar. Up to the period of the American Civil War, there were very few institutional homes for the doing of science, that is, science considered as a research and not as a liberal teaching subject. In the Yale University curriculum the teaching of science was only about 20 per cent of the total pedagogical effort and largely restricted to instruction in elementary subjects. As in the case of Cambridge University before 1850, science was justified in two ways: first, as a form of liberal education, that is, a way of disciplining the mind;[20] and second, as part of the tradition of natural theology inheritied from the Enlightenment. American professors of science no less than British ones were expected to be gentlemen, to be cultivated, to be liberal and not servile, broad but not specialized.[21]

I am not sure how 'cultivated' they were to begin with, but I do suspect rather strong social ties between them and the genteel classes, conceivably stronger ties, strange as it may seem, than existed in the comparable British group. I hypothesize that in Britain the scientific community may have had to emphasize career-building at an earlier age than its American equivalent. Perhaps European competition had much to do with this.

The search for a steady institutional support base for science appears to have emerged several generations later in America than in Britain. Once again I would like to suggest that supply exceeded demand. Americans returning from abroad after study in Germany were imbued with the idea of the advancement of knowledge through discovery. They did not find adequate support in the existing colleges and universities and sought alternatives. Just as in Britain, there was a new universities movement in the United States, but it consisted of both public and private universities. The state universities, largely meant to service developing sectors of the economy like agriculture and mining (with some wonderful and ingenious variations, as at the University of Missouri where a classics professor offered 'agricultural

Greek'), were less ready to adopt a pure research function than wholly new foundations like John Hopkins and the University of Chicago, led by presidents of energy, enterprise and vision. Yet while the technological mission was foremost in the establishment of state universities, an opening was at the same time provided for basic science, since physics, mathematics and chemistry were certainly needed in even the most pronounced technical fields. In some institutions, therefore, technology provided another avenue into the university for research science, although I would not want to minimize the influence of traditional subjects like mathematics which, as the British example of Cambridge shows, was also a means of entry for the physical sciences.

A decisive role in introducing research science into the universities and colleges of America, whether new institutions or old, was undoubtedly played by the American university president who worked closely with boards of trustees. Charles W. Eliot at Harvard, and Daniel C. Gilman at Johns Hopkins were intensely interested in folding science into the university curriculum. Gilman, incidentally, had pushed science while at Berkeley but could not convince potential supporters of its value. James McCosh, a Scot who brought British ideas of liberal education with him to the congenial environment of Princeton, was just as determined as other presidents but against rather than for science. He would not put science under a full head of steam in New Jersey.[22]

The American university president of what has almost come to be known as the 'heroic age' of the president had no counterpart in Britain. In the federal constitutions of Oxford and Cambridge the vice-chancellor's powers were severely curbed by the independence of the constituent colleges, and within the colleges the authority of the heads was limited by governing councils. In the new universities of the provincial cities the academic senates rapidly took effective control of the institutions away from the vice-chancellor and the body of lay trustees. In America the boards of trustees retained their legal and policy-making authority, and presidents continued to act on their behalf. Increasingly, however, most of the routine functions of the university associated with research, the academic programme and the recruitment and promotion of faculty were delegated to departments and senates. This is the situation in existence today. The complexity of the large university and the competitive nature of the research university system with its marketable professors has permitted them to take a greater part in actual decision-making. The project system of awarding external grants has given considerable authority to the professor in his role as principal investigator, so that outside institutional support bypasses the office of the president. However, the recent emphasis on accountability in the award of external grants, as well as affirmative action programmes and the solicitation of private funds have restored to campus administrators some of their former authority, although their direct personal influence is not nearly as remarkable as in the earliest days of institution building.

In the America of the late nineteenth century, rapidly industrializing after the Civil War, there was no immediate demand for pure scientists, although the nation in general had already made a strong commitment to engineering and technical subjects. In the physical sciences a disparity had grown up between young physicists, for example, who were excited by brilliant work in Europe on the atom, on radioactivity and on quantum physics, and the business sector which had no room for them. The expanding university sector was where they had to be absorbed, but by no means did the individual states provide adequate support for pure science. They saw their main task as providing sums for teaching, especially in a period of rapidly rising matriculations.[23] Just as in Britain, however, the changing structure of business and the chemical revolution produced opportunities for the employment of engineers, chemists and physicists. Openings appeared in Western Electric, the manufacturing subsidiary of the American Telegraph and Telephone Company, in DuPont, the giant chemical firm, in the petroleum industry and at General Electric. Even so, stockholders and management were not always quick to see the advantages of expenditure on research that did not initially appear productive, and for many of the new electrical and chemical industries tension between the proponents of pure and applied research was a regular feature of the life of an industrial scientist. Up to the First World War there was still a discernible gulf between the research scientist and industry, and splits began to open up as well between those scientists who were willing to work according to the schedules and requirements of industry and those who laid hold of older traditions and wanted research to be disinterested and pure, and preferred, therefore, appointments in universities where these values were respected.

SCIENCE, GOVERNMENT AND WAR

As science changed its mission and became a professional activity oriented towards original discovery – a development that was long in the making – the problem of adequate support increased. Periodically in both countries, especially in the second half of the nineteenth century, concern had been expressed that the search for patrons would compromise the essential nature of scientific enquiry, which was the advancement of knowledge. This concern, however, could only surface at that point in history when the ideals of scientific investigation had shifted to the pursuit of original knowledge. When science was an unpaid or less capital intensive activity – when a Darwin could equip himself and sail off on the *Beagle* – this was not a problem of the same sort of magnitude, nor was it a problem when science was viewed as a form of liberal instruction or an aid to theology, or when, in universities, professors regarded their proper role as the dissemination but not the advancement of knowledge. When the mission of science changed, when science became the institutionalized quest for knowledge, then the position of scientists with respect to outside support also had to change. And as the doing of science was more costly after 1850 than before, the question of the terms on which funding could be accepted became more problematical and scientists more anxious, especially as the request for support exceeded the supply of it.

The decisive point in the history of the institutionalization of Anglo-American science, if I may be permitted a convenient hyphen, was reached in war. The impact of world wars, defence and imperial expansion in bringing government and science together is of such magnitude that it can hardly be disputed. Awkward as this may appear to be, war did more to provide the institutional backing for modern science than any other single development. This is hardly a surprise, but it requires some discussion nonetheless. Although the main interest of government was the applied usages of science, under the umbrella of the war effort came opportunities for the most varied kinds of basic research: contracts for consulting, directly funded projects, the expansion of government-owned research laboratories and government support for universities. As funding arrangements were worked out, especially in the United States, the more limited or precise government objectives adopted at the outset were exceeded or widened.

The history of government involvement with science goes back, in the case of Britain, to the late seventeeth century, when Britain was extending her imperial possessions largely at the expense of the Dutch, French and Spanish. The most notable collaboration between government and science occurred as a consequence of the founding of the Royal Society, which offered advice on navigation, the steam engine and gunnery. Not a great deal of scholarly attention has been paid to the Society's late eighteenth-century and early nineteenth-century history.

The received opinion is that the Society turned away from applied science and from an active association with the various branches of government. We may have to revise our opinion, however, in light of Marie Boas Hall's recent note in *Isis*. Taking another look at the Society's Council minutes, she has discovered a considerable variety of links between the Society and government ministries when Britain was engaged in a momentous struggle with Napoleonic France. But the ties continued on well past this period. The Board of Ordnance sought advice on gunpowder and cartridges, the Navy Office on copper sheeting and lightning conductors, the Customs Office on the fumigation of bales of cloth. The Treasury consulted the Society on the establishment of overseas observatories, the government sought assistance on the construction of what became the Victorian Embankment along the Thames, and so on. In the midle of the nineteenth century the government decided to broaden its interest somewhat by providing a meagre £1000 per annum for the promotion of science generally. The Royal Society was asked to administer the fund, which went for the support of individual research or for specific investigations.[24]

These examples show us two ways in which government can and has made use of independent associations and their officers and members. The first is as a consultative body. Advice is sought on a number of matters with economic and military application. Recompense is not automatic (no money, in fact, was provided in exchange for the services mentioned above). The second way is 'simulation', wherein one kind of institution behaves in a fashion more characteristic of another. In providing the Royal Society with a small sum, Her Majesty's Government was acting like a philanthropic body that wishes to encourage a particular enterprise, leaving the decision on how best to achieve this to the initial recipient. Market considerations do not enter into the first relationship, but in the second the association, in dispensing its grant, is free to take the market into consideration.

In the nineteenth century there was more direct government involvement with scientific work than is usually realized. The government supported tidal, ordnance and geological surveys, for example. It sponsored research into illuminants and lens systems for lighthouses, an obvious interest for a nation that in 1850 controlled 42 per cent of the world's maritime trade.[25] It had scientific posts at the Assay Office of the Royal Mint, the Observatory at Greenwich and the Botanical Garden at Kew. The Medical Department of the Privy Council underwrote various kinds of scientific projects. The Inland Revenue and Excise Department sponsored research into hydrography, munitions and astronomy. Geological support was channelled through the Museum of Economic Geology which was associated with the Commissioners of Woods and Forests, as was the Mining Records Office. In 1851 a School of Mines and Science Applied to the Arts was established. Parliamentary grants were given to the various royal societies, sometimes as general subventions, sometimes for specific projects. The Board of Agriculture provided grants from the 1890s onwards and was followed by a Development Commission for Agriculture and Fisheries. Teacher training grants certainly stimulated the production of science and mathematics teachers.[26]

The Victorian state could honour certain historic arrangements but was usually reluctant to embark on what appeared to be new directions. Despite the pre-existing alliances between applied science and government, each fresh scientific venture had to be renegotiated and Cabinet or civil service reluctance overcome. An excellent case is the establishment of the National Physical Laboratory around 1900. The idea for the project took wing in a period of great imperial economic and military rivalry. For some decades prominent scientists had been asking for the creation of a national laboratory that could undertake research for industry. There was disagreement among them, however, as to whether the primary mission of the laboratory should be pragmatic, whether it should content itself with testing electrical equipment or analysing chemical compounds and metals, or whether it should be pointed towards basic research. A number of leading scientists believed that only the universities should be entrusted with basic research. In any case, the British government saw no need for an industrial laboratory. Lord Salisbury, who was then Prime Minister, supposed there was some benefit to be gained from the establishment of a testing and standards laboratory, but would go no further. In his marvellous words, 'you will readily admit that research into the secrets of nature affords a horizon to which there is no end or bound.'[27]

Scientists found considerable indifference to basic science in the senior levels of Whitehall, especially at the Treasury, which was required to furnish a modest annual grant towards operating expenses.

The mandarins thought the laboratory should be user-supported and in the long-run totally self-reliant. Support from the laboratory from outside was in fact forthcoming. Fees doubled in the period before the First World War, and small private donations rounded out the budget. Testing was also undertaken for the Admiralty and the War Office. But government interest only really picked up when the laboratory began to undertake aeronautics research, which had clear military application. Treasury grants for this work increased strikingly as war approached.[28]

Until the newly founded Department of Science and Industrial Research took over management of the laboratory in 1918, the Royal Society negotiated with the Treasury for the annual grant. The new arrangements solved the funding problem but also raised fears that science had acquired a new master.[29]

The First World War increased the amount of government assistance to science, but it did not happen at once. A number of prominent British academics complained at the time that the government was in fact slow to take advantage of the skills and talents of university scientists. There were some egregious and tragic blunders. A notable one was the death of the extraordinary physicist, H.G.J. Moseley, who enlisted in the army only to die at Gallipoli, and similar stupidities continued on even to 1917.[29a] Valuable discoveries, such as Chaim Weizmann's process for the manufacture of acetone, essential for naval gunnery, was overlooked until 1916. This was of a piece, however, with the government's failure to mobilize the industrial effort until a year into the war when the shortage of shells provoked a crisis. The turning point came with the establishment of a Ministry of Munitions in 1915 which called upon the universities for direct assistance in fighting the war. University scientists discovered new methods of creating explosives and battlefield gases, developed the gas mask, created new medical drugs, improved the properties of optical glass, essential for gunsights, rangefinders and periscopes – a German and Swiss monopoly before the war – worked on aircraft and the hydrophone for detecting enemy submarines. The list of inventions is too vast for summary. Suffice it to say that innovation went on in every discernible scientific field, and the universities at Cardiff and Liverpool even undertook the manufacture of military hardware.[30]

Sanderson, whose treatment of the theme of war work is illuminating and exemplary, maintains that the ties between university science and industry, which were strengthened during the years of the First World War, proved to be lasting, and he gives several examples of continuing cooperation. However that may be, it is certain from the record that scientists rather suspected that the new alliances both with industry and government might prove too limited. In May 1915 the Royal Society initiated government talks which eventually produced two important innovations with long-lasting consequences for the support of twentieth-century British science. The first was the creation of the University Grants Commission, to place the funding of the university sector on a regular basis, and the second was the Department of Scientific and Industrial Research, to further a wider connection between science and industry. It is perhaps worth noting that the Department still fell short of what distinguished academics like the Nobel Prize winner J.J. Thomson, then President of the Royal Society, and the seminal economist, Alfred Marshall, wanted, which was a science and industry office at ministry level.

Between the wars the Department of Scientific and Industrial Research found that many industries were simply uninterested in the benefits of applied science, and where this was the case, when coaxing and encouragement failed, the Department established its own laboratories. The real difficulty lay in the method by which firms made their needs known to the Department. A given industry was asked to form a number of individual firms into a trade or research association which would then apply to the Department for assistance in solving specific industry-related problems. But many firms not only failed to see the direct benefit to them of such arrangements, they also sensed an erosion of their competitive market position if the results of research were shared widely within the industry; and the large-scale firms especially, as they were able to afford their own laboratories, remained aloof. By contrast, since university internal resources were mainly directed to teaching, the university request for funds from the Department through the associated research councils tended to increase. By 1964 (and probably long before) when it was disbanded following new arrangements for funding basic research and postgraduate training, the Department had shifted its function from a government body with responsibilities to industry to one with exceptionally strong commitments to university research.[31]

In America no organization comparable to the Department of Scientific and Industrial Research was created to stimulate scientific and technological research, even after the centralizing efforts required to manage a global war. America remained committed to a liberal political philosophy long after it was rejected in Britian. The absence of such an organization or agency is typical of a society that has always preferred to disperse incentives and dislikes giving too much power of initiative to a single source. American scientists have prospered in an environment where plural sources of support are available. Projects rejected by one agency can be welcomed by another. Individual Congressmen serving on key appropriations committees can be separately approached, cajoled, flattered and persuaded. When political power is widely shared, the search for research funds may be time-consuming, but in the American experience, the effort has proven uniquely successful. In a number of respects, however, there are crude similarities between the United States and the United Kingdom.

In the USA as in the UK two world wars stimulated a closer association between science, government and industry, and many scientists welcomed this association since they had been relatively unsuccessful earlier in achieving the kind of support they regarded as crucial for big science. Once again government, whether state or federal, was more interested in applied science and technology than in basic science. There is an earlier history of state intervention, if a limited one by comparison. In the nineteenth century the federal government had concerned itself with exploring and developing the vast interior natural resources of the continent. The government supported a weather service, a naval observatory, coast and geodetic surveys, ordnance facilities, a Bureau of Standards and agricultural improvement stations; and these in turn provided jobs for chemists, agronomists, meteorologists, metallurgists, mathematicians, physicists, astronomers, mining and hydraulic engineers, etc. At the turn of the century, when America was rapidly urbanizing, the federal government began to supply technical assistance to manufacturing.

American scientists, just as their British counterparts, had long disagreed on the effect of a closer link with government. The issue had been debated by the American Association for the Advancement of Science in the 1880s. As in the British case, the relative paucity of opportunities for the doing of science in industry ultimately threw the leading scientists towards the universities (at least in physics), and from the universities they approached government. It was believed, not without justification, that as government must pursue the common welfare, it could be persuaded to take the widest possible view of the utility of science. However, there were always scientists in Britain and America, and often the leaders, who were willing to extol the merits of science for the practical benefit of society. How much they believed in their own rhetoric is a matter of judgement; but it may at least be suggested that the issue was not always the utility of science but the terms on which support would be furnished – for example whether immediate results would be demanded – and the degree of control to be exercised from outside the scientific community on the project itself, on the hiring of staff, on the construction of laboratories, on matters that the working scientist wanted to be kept a matter of professional discretion.

Members of the American Association for the Advancement of Science discussed the desirability of a closer involvement with government, but members of the National Academy of Sciences, founded during the Civil War, actively sought it. During the First World War the Academy established another body, called the National Research Council, which was essentially an advisory body occupying a position not unlike that held by the Royal Society. Later a National Academy of Engineering was established in confederation with the Academy, partly to assuage the anger of technologists who had not been invited to join the original Academy. After half a century, the number of different scientific bodies providing advice to the federal government can only be described as staggering. The president has a science advisory council, the Pentagon has one, the army and navy have theirs, the Environmental Protection Agency, the Department of Transportation, the Department of Labour all sponsor science projects conducted in universities. There are quangos like the Atomic Energy Commission, the National Science Foundation, NASA, the National Institutes of Health, which support projects in their own laboratories or fund them in universities. Together these form a decentralized set of institutions, in accordance with American pluralism, with support less concentrated than in the research councils of European countries.

THE PEER REVIEW SYSTEM

In his valuable book, Daniel Greenberg has noted that government support for science in America is based on two principles, that of peer review and that of the project.[32] Peer review actually owes a great deal to the private philanthropic foundations which developed it in the period before the Second World War. The project system derives from the war itself. As part of project costs, universities were granted the right to charge high overheads for administering grants technically not given to them but to principal investigators. For some years the project system was really a device for channelling assistance into the higher education sector when Congress was not yet willing to provide universities with any other kind of assistance. Recently the matter of overhead costs has become a major bone of contention, partly because of public pressure to reduce government spending. Spiralling administrative costs not necessarily related to the funded project itself have led to government enquiries into how the large sums given to universities are actually spent. Improper reporting procedures have been charged. But even within the university communities there is disquiet. Large sums granted to certain academic fields and certain individuals have created disparities in income and status between the disciplines. Summer incomes are available for scientists but not for humanists. Science graduate students can depend upon fellowships while in other disciplines the students struggle to make ends meet, spend considerable time in teaching and delay work on their higher degrees (although this is not the only reason for the delays). The sciences have post-doctoral scholarships; the social sciences have a few; the humanities have none or virtually none. The project system, combined with peer review and the influence of the key advisory committees and elite science associations, has also created disparities between institutions and engendered considerable discontent at universities which regard themselves as worthy but less favoured. The counter argument, of course, is that money ought to go to the best talent and the best organized research units, but have-nots will always suspect that arguments for merit are excuses for privilege.

The peer review system in America, the project system, the establishment of an imposing network of advisory bodies in government, quasi-public funding agencies like the National Science Foundation, the creation of central elite associations, comprise a formidable if hardly invincible system of scientific influence in the United States. They represent the culmination of a drive for professional recognition that began a century ago when science began to sell itself to prospective buyers, and when scientists, like other men of ideas, slowly but steadily learned how to recognize and use market forces. A similar process is discernible in Victorian Britain, but in both countries many scientists turned away from the private sector of the economy when that sector proved, initially, to be resistant to the allures of applied science. Government and the universities were the principal alternatives, therefore, but the universities, fulfilling their historic role of teaching and preparation for the liberal professions, could not provide the resources necessary for twentieth-century science. Ultimately, therefore, science turned to government, and the experience of the wars, followed by global military tension between East and West, expanded and reconstructed the relationship.

SCIENCE AS A PROFESSION

At the beginning of this essay I remarked upon the inchoate nature of scientific communities in the past and the relative absence of hard and fast occupational boundaries between different categories of investigator and practitioner. Energetic but lacking a wholly shared intellectual purpose, scientists were not in a position to pursue singlemindedly their common self-interest. The absence of a clear conception of science as an occupation, however, did not prevent voluntary umbrella associations from arising. These, beginning in the seventeenth century, continued to be founded right into the nineteenth century. As noted, they performed a variety of functions, including the marketing of scientific services and a very limited amount of direct support for research. Nevertheless, scientists could not have improved their support base in more recent periods if they had not successfully learned how to bring influence to bear on modern democratic and bureaucratic government, which, especially in America, is subject to many competing pressures. Scientists were able to do this because, like other professional groups that arose in the nineteenth century, they specialized, concentrated their energies on a clear scientific mission, systematically broadened their institutional means of support and learned how to transform existing associations into reliable lobby groups. They produced leaders, spokesmen and ad hoc committees, all of which, by their very existence, lent credence to the claim that scientists did in fact comprise a community with a single purpose.

Scientists also substantially improved their claim to speak as professional experts. Since for centuries the expert status of scientists was in doubt, or at least ambiguous, this is a point worth further discussion, but to do so it is not necessary to provide examples from earlier than the middle of the nineteenth century.

One of the first steps taken in the direction of professionalism was the drawing of clear lines between career research scientists and the general run of practitioners and dabblers who had for so long populated the scientific universe. Joseph Henry, America's leading physicist in the middle of the nineteenth century, announced in 1846 that 'We are overrun in this country with charlatanism. Our newspapers are filled with puffs of Quackery and every man who can burn phosphorus in oxygen and exhibit a few experiments to a class of young ladies is called a man of science.'[33] In the next few years the emerging scientific researchers competed against the representatives of the tradition of the man of letters for control of the Smithsonian

Institution (the gift of a British philanthropist) and after much Congressional lobbying and shrewd political infighting, they succeeded in capturing the Smithsonian for the science specialist.[33a] In 1874 the research scientists in the American Association for the Advancement of Science assumed the leadership in an organization which, like the British Association for the Advancement of Science, founded in the 1830s, had been an umbrella association composed of amateurs, 'cultivators', technicians, gentlemen of science and all others who had an interest in scientific work, whether or not their careers depended upon it. The amateur or lay members were banished to the periphery. Henceforth only those who were called Fellows could hold office, and most Fellows were career scientists.[34] In Britain the lay element within the Royal Society declined after 1881 as the career scientists tightened their control over that celebrated institution.[35] In sum, a distinction that had been blurred over in centuries past was made precise and associations that had once represented many interests now represented fewer, thus reflecting developments within science itself, where specialization was winning. Some scientists were now 'mere' practitioners, and others were serious, that is, professional. Their lives were dedicated to science, and their stake in it was greater because they were paid. A case might be made for the opposite. Those who are not paid must have the stronger commitment to science for their immediate self-interest is not so great. The dilettante delights (*dilettare*) in his work. No doubt there is some truth in this paradox, but the fact remains that necessity is a primary driving force in human affairs. The dilettante, sufficient unto himself, does not possess the sense of urgency that organizes and institutionalizes thought. Or, as Max Weber phrased it, the dilettante does not know how to estimate and exploit even his most excellent ideas.[36]

Within the universities of America there was a drive to establish science in the traditional curriculum and thereby increase, through a multiplier effect, the influence of science on the nation. Sympathetic college and university presidents appointed and assisted scientists, but perhaps the most significant entering wedge was the American elective system which has no counterpart in Europe that I am aware of, although several new British universities have adapted it to their own requirements. The elective system, which produced the multiplicity of 'majors' now characteristic of American higher education, broke the supremacy of the core curriculum, allowing students a greater opportunity to choose and even design their programmes of study. In Britain diversification of the curriculum was achieved differently. It was accomplished largely through the triumph of specialization, where students usually read only one subject while at university. In America the success of science was greater, for not only did the science teachers have their majors, they also, given the American propensity to believe in the value of general education, had the satisfaction of knowing that all university students took some science as a formal graduation requirement.

The process of professionalism is truly central to the issue we are discussing. For when science turned to government for assistance, and government came to science, it was not merely on any terms that a bargain was struck, as I have already indicated in my discussion of the peer review process. Professionalism improved the negotiating capacity of scientists by strengthening their commitment to the principle of self-regulation and expert knowledge. These principles obtain even where professional scientists are not self-employed. Thus scientists in privately run research laboratories will attempt to simulate the working conditions of universities, where the professor reigns. Where successful, as Kenneth Prandy has shown with respect to Britain, professional scientists and engineers in industry tend to behave more independently than other kinds of employees. They tend not, for example, to join trade unions or follow the industrial model of bargaining relationships.[37] In the civil service, as Karl Mannheim observed, the consequences of a high degree of professionalism is the de-politicization of administration.[38] Generally professions are hostile or ambivalent towards the open market. They prefer to regulate that market by denying the effect of market forces. The clients do not control the relationship; and as if to emphasize that service and profit are quite distinct, Victorian writers on the professions stressed that professionals were not paid for their services, although they might be offered an honorarium. British academics are rather appalled by the operation of market forces within the American system of higher education, which gives a great deal of power to the 'consumer', that is, the students, who are able to shop for courses and professors in the supermarket of the American university. The necessity to please an audience has had a distinct effect on the evolution of teaching styles in the two countries. Even so, the American scholar or scientist also prefers the professional model, and wherever he can follow it, he will. He will try to professionalize the system of securing financial support by making the buyer into a client, by reducing the buyer's influence over the service he is purchasing. Hence the continued appeal to the Pythagorean–Platonic notion of knowledge for its own sake, science for its own sake, all of which implies that knowledge need not and should not be sold on the open market like cattle and

Sonys. The greater the amount of peer review, the more scientists are allowed to disperse money to whomever they think most suitable, the greater their capacity to conduct research along the lines of pure science. Even so, unless their conception of the value of knowledge is similar to that of the universities (as in the case of American foundations at some points in their history), those who supply the money have always been primarily interested in the product. It has been suggested that only a relatively small amount of the research and development money supplied by the federal government is actually for pure research – only 10 per cent of R&D is actually research – and it is also realized that the historic argument on behalf of pure science, namely, that it will one day result in applied science and technology, is largely unproven. At least it is grudgingly conceded that the relationship between pure science and utility is problematical and is perhaps clearest only during the educational process undergone by technologists, who are usually trained in basic science. The conventional wisdom is that technology builds upon itself.[39]

FAUSTIAN BARGAINS?

In his guide to the conduct of the Courtier, the Renaissance writer on manners, Baldesar Castiglione, cautioned the ambitious nobleman not to be too outwardly successful, for he would then surely excite the suspicion and arouse the jealousy of the prince. Instead he was to practise *sprezzatore*. He was to make light of his own achievements, spurn the praise of the world and be gracious to his opponent. If this did not remove, it would at least minimize the risks to himself. Yet we know that the prince read the same book as the courtier and was well acquainted with the dance performed before him. All cultural systems require the maintenance of fictions such as these. Provided they produce no harm, they serve as a means of civilizing conduct and of accomplishing tasks with minimum friction. Pure science has been kept going by large sums granted for the purpose of applying science to a wide variety of social and national uses: environmental research, weapons research, medical research, international economic competition. Anglo–American scientists have negotiated this arrangement as a necessity, as a means of keeping the enterprise of pure science functioning. They have done what medieval astronomers did in order to 'save the appearances': patch up a situation for which there were no clear alternatives. We are all familiar with how saving the appearances is done. Two decades ago every application to the National Institutes of Health emphasized the importance of the project for cancer research. I am told by a friend who is a physicist that he will often, as a peer reviewer, find the phrase 'national interest' used in connection with project applications where the proposed research bears only the slightest relationship to a clearly perceived national interest. In my own area of work I have read innumerable fellowship applications for the National Endowment for the Humanities in Washington which make humanistic claims for the narrowest and most technically conceived historical investigations. Provided the peer review process is intact, the code words are understood.

What has now happened is that scientists have become like Castiglione's courtier. They are too successful. Institutional success is relative and treasury control over the civil service and greater centralization usually means that British scientists have fewer options than American ones,[40] but generally speaking it is accurate to conclude that scientists have built the support systems which they long ago set out to create. They did not do so by themselves, but they were never merely patient observers of larger and impersonal historic forces. That is not the way in which either history or culture works. The support system that obtains in science was built with the assistance and the intervention of scientists who long ago created a culture which in time expressed itself in a network of intersecting institutions. Circumstances and contingencies helped, of course, as they always do. Foremost among them, now as before, *raison d'état.*

The state in western democracies, however, is a very different institution today from what it was in the seventeenth and eighteenth centuries. It is far more formidable and competent, for one thing. Its activities have never been easy to follow in any historical period, but it is even more bewildering today because representative democracies encompass a vast range of plural interests. Recent changes in American politics illustrate the confusion that can result. They have certainly rendered pure science particularly vulnerable.

The decline of political parties in a country with a history of relatively weak parties has produced a vacuum filled by the single-issue pressure group: the environmental lobby, the oil lobby, the women's movement, the animal rights movement, the campaign against the misuse of medicinal drugs, and so on. Each of these has its own priorities for the use of tax monies. Most of them have a public position on the uses to which science should be put and on how university laboratories ought to be managed, regulated and staffed. Furthermore, several controversial issues, such as the nuclear and environmental ones, understandably engage the emotions of scientists as well as laymen, and the resulting division of expert opinion has inevitably weakened public confidence in the judgement of scientists, many of whom are suspected of narrow self-interest and contempt for the public anyway.

Ironically, given current public discontent with the professions, one other source of difficulty for institutionalized science, especially university-based science, is the spread of professionalism into areas of government from which hitherto it has been absent. The very success of the American higher education system has contributed to this. The extraordinary increase in the numbers of young college graduates employed by congressmen, senators and assemblymen at state and national levels has given elected representatives a better command of issues and complexities, or perhaps I should say, has led them to rely more on their staffs for advice than on the outside expert. College graduates, especially recent ones, have a certain amount of inside information on how educational institutions function, at least with respect to matters that concern them, and their perceived dissatisfactions are easily communicated to the political incumbents who rely on them for information on how the peer review system works. (The British MP has nothing remotely approaching this enormous and quite unprecedented staff support expansion. A small step in that direction has recently been taken by the creation of a parliamentary system of standing committees on the congressional model.)

These changes in the functioning of the political system affect output and morale and widen disagreement over research priorities and aims. They also threaten to turn university professional relationships into management–employee ones – a change, should it occur, that strikes at the very heart of the scientist's role and self-conception. This has already occurred wherever the industrial model exists and wherever industry does not simulate or partially simulate the university model of disciplinary autonomy.

The setting of science priorities is also affected by what some analysts have called the emergence of the 'power presidency'. The interest of the executive branch in science matters goes back a long way, but it is essentially an outgrowth of the Progressive Era with its emphasis on social reconstruction and government regulation of the economy. During the 1920s and 1930s, however, physical scientists did not succeed in gaining a firm footing in the White House. Social scientists in the 1930s had more influence in Roosevelt-era New Deal programmes. But since the Second World War presidents have been very much concerned with science and applied science; and, according to one analyst, have acquired a much stronger role in science policy-making.[41] The growth of a huge system of defence contracting and funding for at least 4000 R&D installations has provided an opportunity for the executive branch to exercise greater management of these sectors if it so chooses through its constitutional authority over the bureaucracy which grants the contracts. The executive branch has also acquired some of the glamour that for centuries surrounded the princely courts of Europe where hangers-on lingered in expectation of rewards and recognition, pensions, honours, privileges and favours. Leading scientists instinctively gravitate towards the centre, welcoming appointments as presidential science advisers. All of this increases presidential authority in determining the kind of science that is to be done and in selecting the institutions which are to receive contracts and patronage. It is important to point out, however, that while the presidential office has the capacity to influence science priorities, effective control depends in the last resort on the time and energy White House officials are willing to devote to it.

It is perfectly in keeping with the authority vested in the president to choose advisers who are congenial to him, although it is hoped that he will seek the best advice available. However, it has been suggested that the selection of advisers has become a political or ideological matter, where it is not so much expert opinion but political help that is sought. Issues that lie on the border between science policy and social policy – abortions, for example, or the teaching of Darwinian theory in state schools, issues on which there is currently some division of national or religious opinion – particularly lend themselves to political exploitation. It has also been suggested that recent presidents have openly espoused science programmes which in reality they have no intention of establishing or keeping merely to improve their public image.[42]

SCIENCE AS A COMMUNITY: HOW OPEN?

I need to go back now and conclude with some historical reflections on the notion of science as an open community. Western scientists certainly appear to subscribe to the proposition that freedom of scientific interchange is essential to the progress of science. 'Scientific communication is traditionally open and international in character,' affirms a National Academy of Sciences panel enquiring into the problems of technology transfer in the 1980s (the Corson Report).

'Scientific advance depends on worldwide access to all the prior findings in a field – and, often, in seemingly unrelated fields – and on systematic critical review of findings by the world scientific community. In addition to open international publication, there are many informal types of essential scientific communication, including circulation of prepublication drafts, discussions at scientific meetings, special seminars, and personal communications.'

The panel then goes on to consider the question of national security, concluding that controls on technology transfer should be imposed only as a very last resort, even if risks are involved, and that universities in particular should rarely be subject to limitations on access or communication.[43]

A number of separate but inter-related issues have to be considered. The first is the actual degree to which scientists, practitioners, *savants*, and so on have been willing to share the results of their enquiries. The second is the degree to which, therefore, they have regarded themselves as part of a family, a common culture or a network of learned men devoted to a single purpose. The third is their dislike of secrecy, even self-imposed secrecy, and their resistance to attempts to impose it upon them so that research results cannot be shared widely or across national boundaries.

As we would expect, the historical picture is neither clear nor consistent. Contradictory examples can be furnished. It is very difficult to estimate the extent to which science has been what John Ziman calls public knowledge or has formed itself into invisible colleges. We know little about the international science community as a community before the middle of the seventeenth century when the earliest scientific associations in England and France appeared and later published their transactions and proceedings. A vast amount of information on the creation of invisible colleges lies hidden in private correspondence. Some of it has been made public, such as that of Martin Mersenne and Henry Oldenburg. Much remains to be consulted. It is also difficult to measure the amount of science kept secret in these earlier centuries, either because of the threat of persecution in the Counter-Reformation, or the fear of offending powerful patrons or because of rivalries and jealousies engendered by the race for fame. Personal quirks and special circumstances abound, as in the case of the alchemists, who were very secretive about their work. Much of Newton's writings had to be pried from him. That may have been more the result of uncertainty on his part than a desire to withhold information, and his personal mysticism also enters into the picture. The great seventeenth-century experimentalist, Robert Hooke, wrote down some of his findings in anagram form and deposited them in the

Royal Society to shake off competitors and establish priority of discovery.[44] In the nineteenth century the famous geologist, Charles Lyell, systematically misled the public on the implications of his theories concerning the Noachian Deluge, since the issue was very controversial. At the same time he coaxed into history a view of himself as a man of the utmost probity. He worried desperately about career, appeasing benefactors and achieving public acclaim.[45] As a story of professional rivalry, the desire for prestige, the temporary withholding of critical information in order to ensure priority of publication and fame, we have no book as frank as *The Double Helix*, which is invaluable as an account of some of the contradictions created by the institutionalization of science at present.

We have already noticed that secret research on metals was carried out by London University scientists before the First World War on contract to the steel industry. And of course war-related research has had to be secret. It is understood that where the nation is imperilled, the usual reservations are suspended. That reservations do exist is indicated by the fact that American scientists associated with the National Research Council broke with the military after the First World War precisely because of the secrecy issue.[46]

Resistance to secret research is part of the fascinating story involving the decision made by atomic scientists to withhold publication of information that might have been useful to Germany. Several scientists working in America gradually realized that the discovery of the chain reaction in nuclear fission could result in the development of a terrifying weapon. Leo Szilard, driven out of Hungary by the Nazis, was working on fission with a team of Columbia University physicists in the late 1930s. His imagination foresaw the creation of atomic weapons, and he appealed to leading nuclear physicists, younger researchers, and journal editors to refrain from publishing any information that might be useful to Germany. Many of the principal scientists in Britain and the United States did not believe the danger was imminent. They were disturbed by Szilard's demand for a conspiracy of silence, which a number of them regarded as absolutely antithetical to science. Others went so far as to suspect him of private motives, none flattering. The American government did not see a case for secrecy, but gradually Szilard prevailed. New work was not published. The result was a setback for German atomic development and a corresponding gain for American science.[47]

Wars and international conflicts always complicate issues. The suspension of laws, values and moral scruples that occurs in wars is usually justified as being temporary, as only allowed for the duration of the emergency. Thus the ancient Roman Republic suspended representative government and put war powers into the hands of a dictator. But perhaps it is wishful thinking to believe that war is the abnormal condition as the threat of war continues and preparation for it is undertaken on a colossal scale. The scale upon which modern war is fought places a special burden on the scientists. Hence the issue of classified research continues.

Since in America the universities have a greater share in fundamental research than in other western countries where the research institute is favoured, secrecy is a continuous problem for university-based physical scientists. This is especially the case for those in the University of California system, for they also benefit from research opportunities available to them at government weapons laboratories like Livermore and Los Alamos which are managed by the University of California.[48]

So troubling is the question of classified research in the contemporary university that one hears talk of moving research laboratories into other kinds of institutions where academic scruples may not weigh quite so heavily, or, as a first step, dissociating federally supported laboratories from university management. These proposed solutions are hardly simple. To take only an obvious point, the scientific elite in the United States tends to be concentrated in or near the leading research universities. Diverting public money from university to federally owned and managed laboratories may solve an administrative or ethical problem, but it hardly begins to address the question of the quality of scientific research.[49]

It would not, in any case, solve the ethical problems for everyone, for as has been said before scientists may, at certain times for purposes of institutional necessity, behave as if they were a single community, but at other times the cracks and divisions show. How the individual scientist accepts the restrictions brought about by secrecy imposed from the outside depends upon a number of variables. It depends upon how theoretical or applied his research is and therefore how close to the actual production of destructive weapons he feels it to be. It also depends upon his special field of science, his need for costly equipment, the size of his team and the degree to which his support derives from defence contracts. Doubtless many scientists would not like to debate the question of public knowledge but would prefer to carry on their work untroubled by such academic and perhaps for them irrelevant intrusions.

How the scientist regards secrecy also depends upon non-scientific considerations, such as his personal position on such sticky and indeterminate questions as the national interest, which can (and does) include government controls on the export of technology, the dissemination of technical information and the sharing of research results with foreign graduate students studying in America. Certainly patriotism and strong national feeling have interfered with international scientific cooperation in the past and are not absent today. The First World War, to remark on only one example, virtually destroyed the international community of science as represented by the International Association of Academies, an umbrella organization consisting of twenty-one national academies, fourteen of them being in belligerent nations and four of them in Germany alone. The International Research Council, which was founded after the war as a successor to the International Association of Academies and changed its name in 1931 to the International Council of Scientific Unions, was heavily politicized, national interests continually dominating discussions and programmes. Long before the First World War literary, archaeological, philosophical and geological academies, either of national origin operating abroad or of international constitution, were showing the same effects of patriotic zeal. Many academicians, whether scientists or not, simply accepted their role as instruments of national policy. Their partisanship was no more sophisticated than the crudest jingoism from the classic period of European imperialism.[50]

'Public knowledge' as an ideal faces additional challenges at the moment in America through the combined pressures of a new phase of international economic competition – other nations feel this as well – and the financial stringency affecting so many American research universities. The decline of state and federal support has thrown some institutions of higher education into the arms of private industry. In genetic engineering and biomedical research some professors have actually formed corporations and companies, a matter of concern to Americans (although not to Europeans for some reason) because of their worries about conflict of interest. In 1981 and 1982 the university and scientific establishment in the United States was preoccupied with the negotiations between the Massachusetts Institute of Technology (MIT) and a wealthy industrialist named Edwin C. Whitehead, the largest single stockholder in a corporation that was moving into the field of bio-engineering. In 1982 the Corporation of MIT and Whitehead concluded an arrangement unique in the history of American higher education. Together they established a research institution to be called the Whitehead Institute for Biomedical Research. It was favoured by the MIT administration, and although the opposition to the plan was bitter and vociferous, the MIT faculty approved the venture with some reservations by a ratio of eight to one in a straw vote. There were three main sticking points: Whitehead's insistence on family control of the finances of the new Institute, which gave him some voice in the selection of Institute biologists, who would also hold appointments at MIT; a fear that Whitehead's business interests would lead him to favour research with commercial application; and his demand for ownership of patent rights. Opponents feared loss of control over the selection of research areas and restrictions on the exchange of laboratory results where Whitehead's financial self-interest was paramount. The administration approved of the Institute connection because it increased MIT's biology faculty by one third, provided MIT with operating funds and a huge endowment (when Whitehead died), and put their institution into the forefront of bio-engineering research.[51]

Not being a scientist myself, I cannot predict the effect current trends towards marketing science or strengthening state regulation will have on the cognitive character of science, on the process of discovery and on the pursuit of fundamental research. From the research scientist's point of view, the issue remains adequate support for pure science; but there will always be (at least there always have been) scientists who will feel that even massive intrusions of public money will not serve basic science as well as it will other interests. After the First World War the British mathematician, Hyman Levy, stated that government support of science had a much greater impact on the growth of science as a profession than on basic research.[52] Similar arguments are raised in connection with the National Aeronautics Space Administration and space exploration generally, the pure science advocates expressing disappointment with the way public money tends to be spent when prestige and political considerations are present.

In the past in both Britain and America (and elsewhere) the marketing of science as applied science has often been deplored by some groups of scientists as unworthy of the true purposes of disinterested investigation, as an interference with the search for truth. That is why for them the proper home for pure science has been the university, which, however, has exhibited some of the same vulnerabilities or susceptibilities to the market as other scientific institutions. Some scientists, as well as scholars, have always objected to the practice of assigning a market value to intellectual work. Such objections cannot be easily dismissed. The market as a measure of intellectual worth is inconsistent. Some forms of human activity require a protected environment, although the existence of many competing markets functions as a form of protection. One must recognize the rhetorical dimension of the argument that pure research needs special treatment. It is not merely a question of the importance of one kind of science over another. It is also a question of controls. Scientists, understandably, would rather do research that is relatively free from the kind of external accountability that is associated with applied research and technology precisely because they yield products.

Historically the boundaries between pure and applied science and technology have been imprecise. Significant cross-fertilization has occurred. It has therefore been suggested that the difference between basic and applied science may depend less upon the actual research problem or mission than on such elusive factors as the motivation of the researcher. One of the reasons it is so difficult to estimate the amount of basic research that is contained in an R&D budget is precisely because the personal characteristics of the researcher cannot be taken into account. Because the working differences between applied and basic research are not always clear, money for one can yield results for the other in the normal course of laboratory investigation, no matter where it appears on the balance sheet.

There can be no doubt that historically the demand for science, whether from the open market or from patrons private or public, has been for useful science, for science with economic and technological application. Some scientists have pressed their services on patrons, overcoming the doubts and objections, others have had to be convinced, and still others have resisted involvement in the market in the name of pure science. I am not competent to judge whether regarding science as a product has advanced or hindered theoretical work. Arguments can be found either way. Historically, the question is extraordinarily complicated.

THREATS TO SCIENCE AS A PROFESSION

I have tried to show through discussion and the use of historical examples that from an early date science has had a market dimension, although the preferred option for the doers of science was service in a restricted rather than in an open market, especially in a market where the seller had the upper hand. In time, therefore, science became a profession and acquired a privileged position towards the market. As a profession, science was able to set qualifications for entry into scientific work, influence the job market and through the principles of project and peer review determine both the directions research was to take and the qualifications of individual scientists to undertake it. I have tried to indicate that when we speak about science as an open community we are really speaking about science as a guild, as (in archaic French usage), a *syndicat*, as a self-governing profession that sets standards and establishes rules for the pursuit of those forms of intellectual activity we call 'science'. I have pointed out that pro-fessionalization was the essential historical step taken to solve the ancient problem of adequate support for scientific work, and that the formation of scientists into a professional community is the key to understanding much of the success that scientists have enjoyed in the past three decades.

When we speak about threats to the open community of science we are really speaking about threats to the professional authority of scientists, challenges to the way in which they govern themselves, ask questions, decide the research agenda, identify and reward success and the value they place on the pursuit of knowledge. What is at issue, therefore, is the very culture of science as it exists today, and this inevitably involves the quality of scientific work. Once the authority of the expert is rejected, standards are difficult to establish. There is no arbiter to distinguish the degrees of scientific achievement. We return, in a sense, to that past world of assorted practitioners and purveyors of science, where science was so broadly defined that almost anyone could join in.

How else do the threats we have been discussing affect the nature of the scientific community? To deal with some of the external threats first, how does secrecy affect the scientist's code of professional ethics?

I think it can be demonstrated that even short-term secrecy does have an adverse affect on professional relationships. The peer and project review processes are certainly influenced, for how can hidden work be evaluated? Trust disintegrates rapidly in an environment of secrecy, as even the Szilard episode shows. Existing rivalries are exacerbated, new ones created. For the scientist in industry, if re-quired to limit his communication to only a few colleagues, there are some special problems. Gain and profit inevitably become the modes of evaluation and are substituted for other modes of recognition. It is true that even the most disinterested academic scientist thrives on recognition. Honours, prizes, titles, appointments, speaking en-gagements, a mild amount of media exposure are valued. How could it be otherwise in any form of human activity? But the central animating value of scientific professional life is distinction of mind as recognized by those most qualified to judge; and no amount of power or influence or monetary gain has displaced the esteem of one's colleagues that is the essence of professionalism.

For the university-based scientist the problem is especially acute. The freedom to teach and to learn and to be fearless in the pursuit of truth, to follow the argument withersoever it goes, as the philosopher John Stuart Mill once phrased it, are ideals that have been bitterly contested over the centuries. They are the principal means by which the university has legitimized its elite privileges in the modern world. The ideal of truth in the classroom, in the library and laboratory is difficult to maintain when students suspect their teachers pay only lip service to it.

In general, secret research atomizes and denatures a particular set of social relationships that have been long in the making. In yet another functional way it upsets working relationships based on the team and the research unit. While prototypical instances may be cited, the team is a relatively new form of scientific endeavour, virtually unknown in the earliest days of modern science when the scientist worked more or less alone and kept his own counsel if he so chose. It would be a nice test indeed to try to determine how successful science and applied science are without the cross-fertilization that occurs within the re-search team, without the confirmation and verification necessary for high-level work. Of course whole teams can and do work together secretly. What are the consequences of such arrangements in what may gingerly be called normal times?

In order to understand the threats to public knowledge and to all institutions like universities based squarely on that inestimable idea, we must also be willing to recognize that modern science has often and willingly placed itself in a potentially precarious position. It does not yet approach the position of state dependency reached by science in nineteenth-century France and Germany. Scientists in France in the eighteenth century, especially if allied with the *philosophes*, were quite eager to enlist the authority of the state on behalf of science to break the clerical hold on education of all kinds. In Germany scientists aspired to and enjoyed civil service status as an end to some extent above and beyond science.

Yet Anglo-American science, even with the long and honourable traditions of the self-governing, voluntary society, has entered into partnership with the state. To shift the image, if science will lie down with giants, it must expect them to exercise the wonderful and monstrous power that Miranda warned us giants possess. Scientists should not be surprised if the interests of politicians do not always coincide with theirs, or if they behave ideologically, dismissing scientific advisers according to political belief, or decide to fund projects according to a list of priorities different from what scientists might themselves choose. This may be reprehensible and it may also be damaging to the best long-term interests of the nation – although that question will remain debatable – but it is certainly implicit in the terms of the historic settlement. Institutional arrangements arise from specific historical conditions. They are subject, as Edmund Burke once wrote, to the 'great law of change'. These matters should be understood.

I would like to end this lengthy discussion which has left so much unsaid by referring to another kind of threat to public knowledge and to the institutional arrangements that have been designed for its pursuit and sharing. They are no less than the ambivalences discussed so penetratingly by Robert Merton. They have been a scholarly concern of my own as well.[53] I close with them for emphasis.

Professional groups and university professors are generally more aware of outside threats than they are of inside threats. Professors in particular sometimes act as if they are monks, whose duty it is to serve society in the name of a higher purpose. They can pray but not fight and are therefore powerless. This attitude is understandable. Since professors are not self-employed and have no independent resources, they are dependent upon donors, patrons and benefactors. American professors especially were singularly unequipped to cope with the student demonstrations of the 1960s precisely because of a long history of conceiving of threats to academic freedom, to the special privileges of the teacher and student in the classroom, as always originating from the outside. Even a superficial examination of the history of British and American universities simply shows this not to be the entire story. When Oxford and Cambridge cherished their Anglican connections, heterodox religious views were suspect. Many other examples can be furnished.

The professional ethic of disinterested service has arrived in the present century in wrappers of ancient dignity and nobility. When defending themselves, scholars and scientists are prone to overlook contradictions and dilemmas inherent in their own system of rewards and incentives. The social institutions of science, writes Merton, are malintegrated[54] and I would extend the observation to all of academia. The structure of modern knowledge rests on a late nineteenth-century premise that originality is the leading characteristic of the most advanced and esteemed work. It was not always so. Discovery was not always accorded the special distinction it enjoys today. In the writings of eighteenth-century humanists like Jonathan Swift and Samuel

Johnson, scientists are viewed as madmen who wished to control the processes of nature and deluded themselves into believing that they had a God-like understanding of the nature of the universe. Until late in the nineteenth century originality was thought inappropriate to the education of young persons, who would only be disturbed by wild conclusions and misled into the narrow paths of specialism. Small achievements, it was maintained, would be magnified out of all reasonable proportion, and the resulting puffery of self would damage working relationships within the academic community itself. Research would alienate the undergraduates and adversely affect the relationship between the university and the alumni, whose identification with *alma mater* derived from the memory of happy college days of games, theatricals, beer parties and eccentric professors, at whose tables they once dined, the centre and object of donnish affection. These were not foolish arguments. They had tradition on their side. They were derived from a particular conception of the university and of knowledge, and they were related to very specific educational objectives. Since then a very different kind of intellectual world has developed with a different set of values.

In order to be valued in this new world the scientist must be original, yet what is originality? How is it to be weighed, measured, assessed? At what point can it be recognized? What about the issue of simultaneous discovery? Who gets primary credit in a collaborative effort? How does one cope with the claims of graduate students that professors have appropriated the credit for their laboratory results? What does one do about unconscious plagiarism? Merton cites a marvellous passage from the life of Freud which states the dilemma of originality acutely and ironically. Reading a work on Michelangelo published by an Englishman in 1863, Freud was startled to find there certain observations which he supposed original to himself. He found himself anticipated – surely the most remarkable euphemism in the researcher's vocabulary. Distressed, he consoled himself with two thoughts so typical of all would-be discoverers. The first was that no matter who published first, he himself had made the observation independently. The second was that his own discovery was now 'confirmed'.[55] Any research scientist or scholar experiences the same mixture of fright and reassurance as he attempts psychologically and professionally to cope with the internal conflict between a personal desire for esteem and a sense of belonging to a city-state of common intention, between self-interest and disinterest.[56]

Priority of discovery has been a problem for scientists since at least Galileo, but as a question involving professional ethics it has become more pressing since the knowledge revolution of the nineteenth century when originality was incorporated into the reward system. Contradictions are characteristic of all systems of incentives, as Castiglione made clear (but did little to resolve). Being original and humble at the same time is awkward, as is the withholding of findings when others rush into print, avoiding fashionable projects when funding priorities are tied to them, and sharing knowledge before publication when others may use it first. It is difficult being internationally minded in the face of patriotic appeals. It was the brilliant English mathematician, Charles Babbage, a great propagandist for the importance of research, who as long ago as the 1830s talked about a race between nations with scientific achievement as part of that race. It was Sir Norman Lockyer, the distinguished astronomer, who in the early 1900s asked that British universities be mobilized for imperial struggle with Germany. He wanted two British universities built for every German one.

It is not, I think, the marketability of science that is the only or even the principal issue, nor the desire of powerful industrial states to employ science in the service of national aims. These may be accepted as virtual constants, and scientists have always been in two minds about them. The real issue might well be the capacity of scientists, scientific organizations and universities to maintain a sense of their overarching historical and professional goals in the teeth of pressures and temptations which must, in the very nature of human affairs, be always present if not in the same form or degree.

Ben-David has remarked that the loss of a disinterested ethic for the doing of science will transform the community of scientists into merely another interest group.[57] No discipline, to be sure, can be value-free, but it must be relatively value-free. No body of thought can be wholly objective, but it must be reasonably objective. No professional organization can be without petty rivalries or ideological differences, but it must openly recognize them and minimize their effect on the professional community. No professional can be completely disinterested when in fact his livelihood depends upon a certain amount of self-interest; but it must be self-interest in relation to the service he performs. In a sense, his ambition and aims must be narrow and apparent, limited to that which he has been trained to profess. The legitimacy and autonomy of a profession rest on competence; and competence is especially bounded in the age of the expert. They also rest on the ability of the professional scientist to acquire and preserve the trust and confidence of the publics that ultimately supply him with means to do research. That is particularly difficult to do since the privileged market position demanded by professions is generally suspect in democracies, even as it is also respected.[58]

History does not have an ending. At least it is not possible for the historian to predict the exact institutional changes likely to occur under present global market conditions and international political instability. Market responses will be required. Scientists will make them, at the same time no doubt protesting that the market is a far from perfect mechanism for allocating resources from a professional point of view. The disclaimer is true, but it is hardly, as I have tried to demonstrate, the whole or even the most interesting part of the story of the emergence of science to the commanding position it now occupies in modern societies.

It is therefore fitting to conclude this lengthy discussion with some mention of the social and political traditions which have allowed science to flourish. To speak now only of America, science has been responsive to changing social and economic conditions because it has been forced to recognize the numerous dynamic publics that co-exist in a liberal culture: the press, philanthropic societies, private interests, lobby groups, ad hoc pressure groups, the military, the executive and legislative branches of government, the bureaucracies. All of these have local as well as national forms. The plural environment in which scientists have worked has not allowed them to become a separate or privileged estate, cut off from communication with most of society, highly protected yet highly vulnerable. Free enquiry is best preserved, J.S. Mill argued, when no single part of society – neither government, business, the universities, the learned professions or the Church – is permitted to possess a monopoly of educated talent.

How many masters is it possible to serve? St Matthew said only one. The choice lay between God and Mammon. Modern scientists have served many masters, including God and Mammon. With all due respect to a saint, there is obviously room for a difference of opinion.

ACKNOWLEDGEMENT

In the preparation of this paper I am very greatly indebted to friends and colleagues who generously supplied me with references and materials and discussed facts, ideas and conceptions with me. They have saved me from many errors, if not from all. They are Daniel Kevles of the California Institute of Technology, Berni Alder of the Lawrence Livermore Laboratory and Berkeley colleagues Bruce Wheaton, Martin Trow, Roger Hahn and John Heilbron.

REFERENCES

1. Ross Sydney 1962 Scientist: the Story of a Word *Annals of Science*, **18** (June): 65–85; Miller Howard S. 1970 *Dollars for Research: Science and Its Patrons in Nineteenth-Century America*. Seattle & London, p vii.

2. Ross Sydney 1962 *op cit*, p 73n; Hahn Roger 1975 Scientific Research as an Occupation in Eighteenth-Century Paris *Minerva* **13** (Winter): 503.

3. Schaffer Simon 1983 Natural Philosophy and Public Spectacle in the Eighteenth Century *History of Science* **21** (March): 31–2.

4. *Ibid*, p 27. See also Rothblatt Sheldon 1976 *Tradition and Change in English Liberal Education, an Essay in History and Culture*. London, chapters 4 & 6.

5. Rothblatt Sheldon 1976 chapter 7.

6. Heilbron John L. 1982 *Elements of Early Modern Physics*. Berkeley, Los Angeles & London, chapter 2.

7. Heilbron John L. 1982 *op cit*, pp 157–8.

8. These examples have been given to me by Heilbron.

9. Donovan Arthur 1975 British Chemistry and the Concept of Science in the Eighteenth Century *Albion* **7** (Summer) 131–44; and Cantor G.N. 1975 The Academy of Physics at Edinburgh, 1797–1800 *Social Studies of Science* **5** (May): 109–34.

10. Berman Morris 1972 The Early Years of the Royal Institution, 1799–1810: A Re-evaluation *Science Studies* **2**: 205–40.

11. Donovan Arthur 1975 *ibid*.

12. Heilbron John L. 1982 *op cit*, p 154.

13. Swift Jonathan 1970 *Gulliver's Travels*, New York, p 153.

14. Heilbron John L. 1982 pp 35–6, 112–3, 146–7.

15. Even within the periphery, however, there are faculties who model their professional identity on the elite university sector and are therefore in conflict with administrators who see their first duty as service to the consumer. Certain elite sections of establishment universities, schools of business, for example, are more responsive to the market than the traditional or 'autonomous' disciplines, possibly because they have a well-defined clientèle in mind. (My thoughts on such questions have been greatly aided by discussion with Martin Trow.)

16. A subject of special interest to Harvey W. Becher of Northern Arizona State University, Flagstaff.

17. Rothblatt Sheldon 1983 The Diversification of Higher Education in England. In *The Transformation of Higher Learning, 1860–1930* ed. Konrad H. Jarausch (ed). Stuttgart and Chicago, pp 135–6.

18. Sanderson Michael 1978 The Professor as Industrial Consultant: Oliver Arnold and the British Steel Industry, 1900–1914 *The Economic History Review* **XXXI** pp 585–600.

19. Sviedrys Romualdas 1976 Physical Laboratories in Britain *Historical Studies in the Physical Sciences* **7**: 435, and information from Heilbron.

20. A nineteenth-century shift, nevertheless, from earlier versions of liberal education which were more ethological in nature.

21. Kevles Daniel J. 1978 *The Physicists*. New York, p 21, 26.

22. *Ibid*, p 20, 35.

23. But did the number of hours spent in teaching or in preparation for teaching rise or fall? The same question must be asked about

administrative assignments. These are patterns worth tracing if we are to understand how the research function penetrated the universities.

24. Hall Marie Boas 1981 Public Science in Britain: The Role of the Royal Society *Isis* **72**: 627–29.

25. MacLeod Roy M. 1969 Science and Government in Victorian England: Lighthouse Illumination and the Board of Trade, 1866–1886 *Isis* **60** (Spring): 7.

26. MacLeod Roy M. 1976 Science and the Treasury: Principles, Personalities and Policies, 1870–1885. In *The Patronage of Science in the Nineteenth Century* G.L.E. Turner (ed), Leyden; and MacLeod Roy M. 1972 Resources of Science in Victorian England: The Endowment of Science Movement, 1868–1900. In *Science and Society, 1600–1900* Peter Mathias (ed). Cambridge, pp 111–66; Brock W.H. The Spectrum of Science Patronage. In Turner, *ibid.*

27. Moseley Russell 1978 The Origins and Early Years of the National Physical Laboratory: A Chapter in the Pre-history of British Science Policy *Minerva* **XVI** (Summer): 227.

28. *Ibid*, pp 249–50.

29. *Ibid*, p 245.

29a. None of the First World War belligerents had a monopoly on idiocy. Science students, young researchers and professors from France and Germany also rushed to the front or were mobilized. They served in a variety of combat roles, from infantry to the technical services. Casualties were horrific. It took a year or more for the several governments to understand the effect of such wastage on the war effort. (Information from an unpublished paper by Heilbron John Physicists in World War I.)

30. Sanderson Michael 1972 *The Universities and British Industry, 1850–1970*. London, p 232, and chapter 8.

31. Varcoe Ian 1974 *Organizing for Science in Britain*. Oxford, pp 80–1.

32. Greenberg Daniel S. 1967 *The Politics of Pure Science*. New York, pp 16–18.

33. Miller Howard S. 1970, *op cit*, p 7.

33a Miller Howard S. 1970, *op cit*, pp 15–23.

34. Kevles Daniel J. 1978, pp 41–2. Separatist feeling had emerged even at the time of the founding of the British Association for the Advancement of Science. '[I]t would be desirable,' wrote Whewell, 'in some way to avoid the crowd of lay members whose names stand on the lists of the Royal Society.' The inclusionists won, however, at least initially. See Orange A.D. 1971 The British Association for the Advancement of Science: The Provincial Background *Science Studies* **1**: 326, 315–29.

35. Cardwell D.S.L. 1957 *The Organization of Science in England*. London, p 176.

36. Weber Max 1970 Science as a Vocation. In *From Max Weber* H.H. Gerth and C. Wright Mills (trans and ed). London, p 136.

37. Prandy Kenneth 1965 *Professional Employees: A Study of Scientists and Engineers*. London, p 41, 44, 175–8.

38. Mannheim Karl *Ideology and Utopia*. New Directions, New York, pp 153–64.

39. Greenberg Daniel S. 1967, *op cit*, pp 31–2. It is often impossible to distinguish between the categories. A recent Pentagon listing of basic research projects included 'self-contained munitions', 'electro-optic

counter-measures', 'low speed takeoff and landing' (*New York Times* 26 June 1983). Yet no one disagrees that development leads research. In Sweden R&D in industry means only about 3 per cent for "R." Lindqvist Svante 1981 in *Science, Technology and Society in the Time of Alfred Nobel* Carl Gustaf Bernhard, Elisabeth Crawford and Per Sörbom (eds). Nobel Symposium 52, 17–22 August, pp 300–1.

40. For a role similar to the Treasury's now being assumed by the Office of Management and Budget in Washington, see Katz James Everett 1978 *Presidential Politics and Science Policy*. New York and London, pp 66–83.

41. Katz for most of what follows.

42. *Ibid*, p 215.

43. *Scientific Communication and National Security*, a report prepared by the Panel of Scientific Communication and National Security, Committee on Science, Engineering and Public Policy, National Academy of Sciences (1982), 2, 48–51. The panel's guidelines for deciding when restrictions on university research are warranted have not been easy to interpret, however. See *New York Times* 26 June 1983, EY9.

44. HH. Information from Hahn and Heilbron

45. Porter Roy 1982 Charles Lyell: The Public and Private Faces of Science *Janus*: 29–50.

46. Kevles Daniel J. 1978 *op cit*, p 148.

47. Weart Spencer R. 1976 Scientists with a Secret *Physics Today* (February): 23–30.

48. Defenders of the university's management of the laboratories answer that only a small proportion of the research output is in fact classified.

49. Some writers have even suggested that teaching, professional education and research functions are not comfortably integrated in most systems of mass higher education, or at best only integrated in isolated parts of a given institution. They suggest that the natural home for scientific research is not necessarily the modern university and that for developing countries especially a parting of the ways would be beneficial to both. See the discussion in Schwartzman Simon The Focus on Scientific Activity. In *Perspectives on Higher Education* Burton R. Clark (ed). University of California Press, forthcoming.

50. Schröder Brigitte 1966 Caractéristiques des relations scientifiques internationales, 1870–1914 *Cahiers d'histoire mondiale* **10**: 161–77.

51. See *Science* **214** (23 October1981): 416–7, and *Nature* **294** (26 November 1981): 297. Dorothy Zinberg of the Center for Science and International Affairs at Harvard University very kindly sent me information on the establishment of the Whitehead Institute for Biomedical Research.

52. MacLeod Roy and Kay 1979 The Contradictions of Professionalism: Scientists, Trade Unionism and the First World War *Social Studies of Science* **9**: 2.

53. Rothblatt Sheldon *op cit*. chapters 13 and 14; Merton Robert 1976 *Sociological Ambivalence and Other Essays*. New York, p 32–65.

54. Merton Robert 1976 *ibid*, p 36.

55. *Ibid*, p 42.

56. Science makes discoveries. Are these qualitatively similar to those which historians, literary critics, linguists or economists make? The building-block characteristics of science which make incremental discovery so central – perhaps thereby increasing the professional pressure on young

researchers – may not really be matched in the social sciences and humanities. Fear of the 'scoop' certainly exists outside the sciences. Individual historians certainly are worried that someone else may examine a particular body of sources first and therefore publish first. But other factors strongly enter into the evaluation of non-scientific work: rhetorical questions of composition, art, communication; broad humane questions of vision, wisdom, a grasp of the whole. These are not, of course, absent from scientific work, but are they so paramount?

57. Ben-David Joseph 1971 *The Scientist's Role in Society*. Engelwood Cliffs, New Jersey, p 180.

58. I am, I suppose, less confident than Edward Shils about the internal strength of science's devotion to the pursuit of truth and excellence, as portrayed, for example, in his *The Torment of Secrecy* (New York, 1956). He underplays, I believe, the internal vulnerability of institutionalized intellectual life that allows outside threats to become effective. But I do agree that without ideals and an appeal to the best parts of a tradition the enterprise is forsaken. The scientist's responsibility to society, professional ethics and the role of the public in deciding which projects should be funded are discussed at length in Wulff Keith M. 1979 *Regulation of Scientific Inquiry*. Washington DC.

PART TWO

Science in Transition:
A New Pattern of Interests

CHAPTER THREE
Science as a Commodity – Policy Changes, Issues and Threats

Jean-Jacques Salomon

Science has come to play a critical role as a contributor to the economy only at a recent point in history. In the first stages of the industrial revolution, the pursuit of knowledge could still be accorded a status independent of material concerns. The most materialistic conception of science, that of Marx who explained the development of natural sciences in relation to the development of industry, commerce and trade, still recognized that scientific knowledge, as opposed to technical invention, was a free good similar to the forces available in nature. It costs nothing, such as 'physical forces, like steam, water, etc., when appropriated to productive processess (. . .). Once discovered, the law of the deviation of the magnetic needle in the field of an electric current, or the law of the magnetisation of iron, around which an electric current circulates, costs never a penny'.[1]

Science as a commodity? Only half a century ago, such a question would have appeared as a provocation; and a century ago, as an aberration. Yet, by the same token that air, water and even noise have been incorporated into our economic accounting, science has ceased to be considered as a system of available knowledge 'outside the economic circuit'. Nothing could better illustrate this change than the book published by Renan in 1890, almost a century ago, *L'avenir de la science*. It contained the most fervent pleas in favour of public support for scientific research. The argument relied on was not the potential usefulness of discoveries, it was the intrinisic worth of science itself which demanded public support. The state owed researchers subsidies as an absolute duty, a categorical imperative resulting from the fact that science had ousted religion from first place. But such a duty conferred no right of oversight or regulation, 'any more than subsidies to religion give the state the right to lay down the articles of faith. In one sense, the state can do even less with science than it can with religions; it can at least impose certain internal police regulations on religion, whereas it can do nothing, absolutely nothing, with science'.[2]

Times have changed: the practice of science is no longer amateur or purely academic, but a profession among others exercised around the world by millions of researchers who are not solely *savants* in the nineteenth-century sense, but also engineers and technicians. And scientific research no longer stops at the doors of the university, it is practised in industrial and governmental laboratories whose objectives are not the advancement of knowledge, but industrial, economic or military applications. By the same token, the relation between the 'scientific estate' and governments has changed: public support is no longer without its counterpart. As research activities involve important investments, policy-making bodies expect results which are conceived of by the scientists as external to the aims of science. General de Gaulle is said to have once welcomed the Minister he had charged with scientific affairs by saying: 'So my researchers, you find?' The equation of science with power has become one of the ingredients of the art of governing.

The notion of research and development reflects the measurement of scientific and technical activities as they are accounted on the basis of statistics of the financing of laboratories or the numbers of researchers; but it also mirrors the national and competitive character of policies whose object is to secure for the nations concerned a more direct contribution by science and technology to their power and influence, their competitiveness from an economic standpoint, their independence at the political and military level. The future of science is no longer solely determined, in the minds of the scientists themselves, by the internal requirements of scientific progress; it is closely dependent on political institutions and policy-makers, as well as on the economic, social or strategic considerations of the moment. Basic research – pure science – is sensitive to decisions that reflect the fluctuations of history and circumstances, the economic opportunities and constraints, as well as the swings between international détente and crisis.

The successful instrumental character of modern science has even turned its theoretical and fundamental aspects into an economic and political process of public negotiation, orientation and regulation. Just as the frontier between science and technology is blurred, the frontier between knowledge as a vocation and science as a means for power has eroded. It is in this sense that science has become a commodity whose value is related in economic terms to the laws of supply and demand and in political terms to the objectives and needs of governments. The evolution of science and technology policies since the end of World War II, with their ups and downs in shifting priorities, illustrates clearly the extent to which such a status is a function of the economic, political and military context.

SCIENCE AND TECHNOLOGY IN POLICY 1945–73

Since the end of the Second World War, science and technology policies have changed with the international economic and strategic context. By 1967–68 they were already at a turning point. Investment in research had grown continuously, with three strategic sectors – defence, atomic energy and space research – taking the lion's share. From 1945 to 1965, in most industrialized countries, growth of research budgets exceeded (by far) that of national income, to such an extent that some observers suggested, not without irony, that if it continued national budgets would soon be exclusively devoted to science with the entire labour force composed solely of researchers.

The series of disturbances and questionings that occurred in the late 1960s, such as the student and university revolts, the demonstrations against the Vietnam War and the first ecologist crusades, marked the end of the age of plenty for science and technology. Hitherto, any research programme was considered to deserve support and generally obtained it for the simple reason that it promised to open up new frontiers or to lead to innovations. All scientific research sectors, including the social sciences, received the same advantages that the authorities had formerly reserved for the 'hard-core' sectors such as physics, chemistry and engineering which had proved their worth through their technological exploits.

The majority of those in government advisory circles were physicists, and in order to attract the attention and favour of the authorities, every research project needed to be patterned on the mode of organization and financing of the large-scale technology programmes or to be otherwise 'sold' by the science lobbies. Whatever the field of research and however distant from any short- or medium-term application (eg high energy physics), it was axiomatic to advertise the multiple economic, social and political advantages that society would draw, as though by divine right, from the increased investments.

Ahead of the deadline fixed by President Kennedy, the wager of the National Aeronautics and Space Administration (NASA) had been won – 'a small step for one man, a giant leap for mankind' – but the spectacular success of the Apollo project ended in disenchantment. Instead of going to the moon, surely the same effort could be better expended in trying to solve the problems on Earth. The same disenchantment reached out to most science and technology activities, which were seen to entail not only benefits, but also adverse or unexpected effects on the environment: for example, side effects of excessive economic growth itself, or the threats presented by the nuclear megatons housed in the warheads of the tens of thousands of missiles standing ready.

In the wake of the debate on the limits to growth, the indictment of the military-industrial complex, the counter-culture movement and the criticism – even within the scientific community – of the directions taken by research, a new climate developed in which optimism about scientific and technical progress, which since the eighteenth century had been seen as the guarantor of social (if not moral) progress, gave way to doubt and confusion.

It was then that the predominance of the physicists, those of the generation that had taken part in pre-war nuclear research and the development of the first atomic weapons, began to decline in government advisory committees in favour of representatives of other disciplines, such as biologists, social scientists and above all economists. It was also at this time that the objectives of science and technology policy that had been defined in the context of the cold war and the arms race began to be revised. Large-scale technology programmes based solely on considerations of strategy or prestige declined in importance in favour of sectoral policies with more focus on economic repercussions for the civilian sector. In the United States the abandonment of the SST (Concorde's rival) and the Mansfield amendment prohibiting government agencies from subsidizing research not related to their objectives, and in the United Kingdom the Rothschild customer–contractor principle (no public support without a potential customer), confirmed this reversal of doctrine which most industrialized countries were to adopt in declaring the need to take economic and social needs into account and to control technical change more effectively. Similar sentiments were to be found in the Brooks Report, *Science, Growth and Society*, issued by OECD in 1971.

Looking back, it can be seen that the 1967–68 changes and disruptions in research and innovation systems were harbingers of the storms that were to rock the world economy in the 1970s. In the most industrialized countries, the concern with 'relevance' and efficiency was all the less acceptable to the scientific community as the funds allocated to fundamental research ceased to increase or even decreased. Universities criticized, not without reason, these see-saw policies which destabilized research teams and programmes. They blamed the authorities on the one hand for not doing enough to support science as distinct from technology, and on the other hand for doing too much and trying to subject science to criteria of social utility that were not its concern. The uneasiness felt in scientific circles was all the greater since the universities had suffered from the impact of a sharp increase in student population, leading to rapid and often uncoordinated changes and resulting in structural maladjustment and a new imbalance between staff and students, education and research. Stationary levels of funding prevented the renewal and mobility of researchers, with the risk of aging and obsolescent teams from which the universities might take several generations to recover.

CONSEQUENCES OF THE 1973 OIL CRISIS

The 1973 oil crisis broke in the midst of all this rethinking, questioning and uneasiness about scientific researchers. This did not so much cause as accelerate and accentuate the economic and social difficulties that once again transformed the background against which science and technological policy had to be defined. Rapid economic growth facilitated by cheap energy and by a pattern of production and consumption that paid little heed to wastage was succeeded by the simultaneous constraints of far more expensive energy, flagging growth and above all the combination of inflation and unemployment.

Needless to say, the research and innovation systems were not immune to these changes and pressures. In 1973 all industrialized countries had shifted their research effort to the energy field (beginning by concentrating on nuclear energy and then gradually rediscovering the potentialities of older sources such as coal or discovering new ones such as solar energy). But over and above the spectacular increase in the price of oil, they discovered that the difficulties facing them were more deep-seated and required a complete overhaul of scientific and technological research strategies.

The transfer of traditional industries to the developing countries, the ability of an increasing number of these countries to enter world

narkets and compete with the manufactures of the most advanced
ndustrial countries, the dual phenomenon of the irreversible inter-
nationalization of industrial trade and the inevitable specialization of
world markets altered the allocation of the cards to such an extent that
it seemed as though the rules of the game (and even the game itself)
had changed. Science and technology policy was suddenly obliged to
come out of the retreat into which the confusion and questioning of
the late 1960s had driven it.

As scientific and technological research was clearly the greatest asset
of the most industrialized countries in their competition with the
newly industrializing countries, the problem of technology transfer
became the key issue in North–South negotiations. Caught as they
were between the requirements of competitiveness and foreign trade
and the simultaneous pressures of inflation and unemployment, most
countries found that it was not easy to identify new great technological adventures nor above all, to devise miracle treatments to
stimulate technological innovation more effectively and modernize
the productive system. Nevertheless, to pay the oil bill it was essential
to export and therefore to innovate and introduce the technological
changes from which would emerge a new mode of development for
the future.

Compared with the 1960s, when the research drive was sustained at
the same pace on nearly all fronts and when product innovation
outweighed process innovation, the trend seemed to have reversed. In
many sectors and countries the tendency was to focus on low-risk
projects promising rapid returns. Process innovation thus took precedence over product innovation. Industrialists complained of excessive
controls and of the bureaucratic and technocratic suspicion embodied
in the regulations that had proliferated during the previous decade in
the name of environmental protection and human safety. It was true
that a number of disasters – from Minimata to Seveso – had justified
these precautions, but the price to be paid for the sake of the community was the reduction or loss of innovations (for example, in
pharmaceuticals).

Moreover, between one sector or branch and another the research
effort is now very unequally distributed. Apart from electronics, bio-
engineering, and machine-tools, where new breakthroughs are continually taking place, a number of sectors previously considered as
being highly innovative (particularly chemicals) now seem to be
marking time or to have lost their impetus. The application of R&D
results and the possibility of innovation depend very much on industry's ability to invest or on its access to funding. But this is

precisely what appears to be lacking because of the crisis, particularly where research is concerned. The latter seems to be less profitable in many cases than it was ten years ago, and this leads some firms to sacrifice their long-term projects for activities which can be quite unconnected with research but which are likely to bear quick profits.[3]

None of these considerable changes is more revealing than the concern of some American business circles over competition from Japan and even from Europe. This also illustrates the relativity of the controversies that can mark and inflame a period. In the mid-1960s, Europe felt so left behind in the science and technology field that it seemed as though the American challenge could never be taken up, but by the end of the 1970s, the trend seemed to have reversed. In the United States, many observers and authors were complaining of the technological gap which now separated their country from Japan, Germany and even Europe as a whole. For some years now, reports issued by the National Science Foundation (the *Science Indicators*) have suggested that in many sectors American businessmen are less aggressive where innovation is concerned than their Japanese or German counterparts. This apparent loss of technological leadership by the United States would seem to be accompanied by a general slowing of growth in productivity.

Whatever the signs of diminishing momentum, however, it would be wrong to conclude that there has been any decline in science and technology in the United States. With its vast land area, immense resources and its R&D investments which, in absolute figures, far surpass total European investments, together with its unequalled industrial and university resources, the United States remains the greatest world power where research and innovation are concerned. But it is true at the same time that in the key non-defence sectors where international competition is becoming increasingly keen, other countries have scored a number of successes. These signify not so much a waning of US power in absolute terms, but rather the accession to world power status of other countries that had previously been left on the sidelines in the aftermath of the Second World War. Thus Japan and the Federal Republic of Germany have now overtaken the United States in several areas of innovation with industry-wide applications.

Furthermore, the indirect influence of military R&D programmes on technological innovation in the civilian sector should not be underestimated. The size of the Reagan administration's defence R&D budget is eloquent: $24.5 billion in 1983, or more than half the public funds allocated to R&D (nearly $43 billion). With a military R&D

budget increasing by over 21 per cent, there will inevitably be spinoffs from the programmes.

The technologies developed in the nuclear, aerospace and electronic fields in order to perfect the various arms systems have an impact on the civilian sector in the form of new products and processes which, though not directly reflected in increased productivity, nevertheless contribute to the strength of American industry. It is hard to deny the influence of defence research on the wave of innovations that led to the spread of electronics, computers and even robotics in the 1950s and 1960s. The overall effect of public R&D investment on the economy would perhaps be greater if it were made directly in the civilian sector, for industries competing in the consumer-goods market. But could this happen in the United States if defence aims were not there to spur the administration to action?

A TIME OF UNCERTAINTY

Technical progress cannot be taken for granted, any more than can economic growth. Under the pressure of the constraints and changes arising from today's new context, the research and innovation systems must adapt. They are being subjected to new demands and need to explore new opportunities and outlets. For example, competition with the newly industrializing countries is compelling OECD countries to introduce a process of continuous change in their output patterns, in order to replace the more traditional products that developing countries are beginning to manufacture for themselves or for export. 'In this sense, intellectual captial – scientific resources and the aptitude for technological innovation – constitutes the major asset of industrialized countries in the new modes of international competition and interdependence.'[4]

In all OECD countries, industrial structure was previously based on a certain pattern of relative factor costs, but that was turned inside out by oil price rises. It has become essential to restructure industry in line with the new pattern of factor costs. Because of these new energy-market constraints, it is important to save energy, recycle waste products and speed up the development of new energy sources. Similarly, innovations are required in the service sector in order to meet society's new demands, particularly as regards technologies which will improve living and working conditions. This is an area that has been left on the sidelines by the technical revolutions that have transformed agriculture and the manufacturing industries.

Without rapid technical progress there can be no economic expansion, and vice versa. In these crisis years of the 1980s it is obvious that a structural adjustment of the industrialized economies is wholly dependent on technology, both as a guarantee of renewed growth and as an insurance for the future. But the great paradox of the current situation with regard to research policies is that while there has never been greater awareness of their economic and social implications for the future, never has it been so difficult to find the investment needed to meet the many challenges inherent in those implications. Science and technology policies today are more than ever dictated by economic and foreign policy considerations. But these short-term pressures are such that the research activities on which future growth depends are liable to be starved of funds.

A vast restructuring movement is now taking place in industry and services, social practices and individual behaviour patterns. Scientific and technical progress – the heart of this movement – is an indicator of the promises and risks of the changes to come. We know only too well where tomorrow's fields of action lie and we can glimpse some of the consequences that the new discoveries and inventions will have on our work, our life and our leisure. However, it is essential to at least preserve (if not develop) a climate which is favourable to scientific research and development and this is what is at stake today.

The greatest risk to which any research policy is exposed nowadays is the following vicious circle: no economic growth hence no investment in research and innovation hence no economic growth. However, as scientific research and technological innovation depend largely on state support and incentives, most countries are now obliged to restrict public expenditure. The welfare state has also been challenged by objections concerning the use of public money for venture capital. Consequently, the economic and social context restricts the field of initiatives which contains, along with the future of research, the chances of renewed growth for the industrialized societies and thus the survival of the developing countries.

The oil market and the world energy market have become politicized. Prices and production are influenced as much by political considerations as by the play of market forces, if not more so. The search for alternative energy sources continues in parallel with a research drive focused on traditional sources and nuclear power stations. But even if the oil price diminishes, the threat of blackmail will still exist.

Not only does the new distribution of world economic and industrial power increase the number of competing countries but it also leads to innovation policies directed at the same technology goals.

Within the OECD area, the role of 'locomotive' is no longer played by the United States alone, but is shared between the latter, Japan and Western Europe. Outside the OECD area, some developing countries that have reached the industrialization stage are playing an increasingly important role on the world market. It is inevitable that in this harsh struggle, which some already see as the lead-up to an all-out economic war, some will be victorious and others will be vanquished.

In the economies of industrialized countries, social goods and services are becoming increasingly important. As neither the supply of nor the demand for these services is subject to the traditional laws of the market economy, their production and consumption represent increasingly heavy overheads to be borne by private goods and services.

Where employment is concerned, the service sector in general is now playing a role similar to that played by industry relative to agriculture during the period of rapid industrialization. Consequently, given agriculture and industry's ability to produce more without creating new jobs, the question arises as to whether the service sector, on which the rapid introduction of information technology has a particularly strong impact, will be able to create more jobs than it eliminates.

Lastly, mention should be made of the emergence of new social values and aspirations. The importance attached to environmental protection and the quality of life, the growing weight of social goods and services in aggregate demand, the changing attitudes with regard to work, and last but not least, the more critical stance of the general public on science and technology, all these have an obvious influence on future research and development.[5]

This complex of changes delineates a world in transition, more uncertain than ever, where the long-term prospects are difficult to identify and were medium-term analysis is inadequate for the purposes of coherent strategy-making. Inflation, declining returns on capital and the persistence of high interest rates result in low private investment and thus short-term concerns are liable to take precedence over medium- and long-term needs. Yet the whole point of a research policy is to pave the way for the future: there can be no instant results because of the length of the transition from discovery and invention to development and innovation. Technical progress is the guarantee that new growth sectors will be created, but during a period of stagnation, it is becoming increasingly difficult to take this essential response to change for granted.

THE TECHNOLOGY IMPERATIVE

It is a sign of the new concerns and priorities, that the institutional model for the agencies that draw up and implement science and technology policy has changed radically. Since the end of the 1950s it had been based on that in the United States, namely a lightweight structure situated as closely as possible to the supreme executive power, without any direct responsibility for budget management. The Office of Science and Technology was thus set up at the White House under President Eisenhower in order to obtain advice from scientists, estimate research potential, identify inadequacies and needs, stimulate pilot projects, and if possible coordinate the national R&D effort.

This model was more or less directly adopted by all OECD countries as they gradually came to recognize the importance of science and technology issues in the present management of public affairs, at both national and international levels. Thus in France the *Délégation Générale à la Recherche Scientifique et Technique* (DGRST), headed sometimes by a minister and sometimes by a state secretary delegated by the Prime Minister, was mainly concerned with information, consultation and coordination, having no real management responsibilities apart from the so-called 'concerted actions'.

'How can we become like the Americans?' wondered the European scientists who, in the aftermath of World War Two, envied the research structures and potential of the United States, their ability to rally the scientific community to large-scale projects, and the number of their Nobel prizewinners. 'How can we become like the Japanese?' is the question asked today by American and European industrialists who envy the vitality of Japanese firms, their ability to take advantage of advanced technologies and their winning strategies on international markets. The institutional model has moved from Washington to Tokyo. It is now the Japanese Ministry of International Trade and Industry (MITI) that serves as the point of reference, a huge inter-sectoral department closely associating R&D, industrial policy and foreign trade under the same administrative supervision. Thus, a Ministry for Research and Industry was set up in France in 1982 inspired partly by this example.

Whatever the model, there is no doubt that the prestige conferred by the award of a Nobel prize in the Science Olympics or by the success of a major national project like the Apollo moon-landing is of less importance to the politicians than the added value of the new products launched to conquer international markets. Science and technology policy stands, from the outset, on the most sensitive front of the economic war; namely, where the R&D effort is translated into industrial modernity and exports. MITI is so much a part of Japan's success story that some people are close to seeing its structures as the panacea for the modern state. This comment is an illustration: 'In each century, supremacy rests with those nations that first discover the political conception appropriate to their time. To paraphrase Tocqueville's observation in more pragmatic language, one would wonder whether a nation has given itself a structure equal to its strategy'.[6]

The fact is that the future of research depends more than ever on resolute and coherent policies of which it is easy to see what the central aims must be: namely, to preserve and develop an efficient scientific and technical research system not only in the universities but also, and above all, in industry and the service sector. We know that technical progress results from a combination of micro- and macro-economic factors – physical, institutional and organizational – which is still by no means certain to arise. 'Maximising the chances that such a combination will arise is precisely the objective of government innovation policies.'[7] To pump in resources is not sufficient; it is also necessary to create the appropriate state of mind and conditions, which means adjusting the structures in which industrial entrepreneurs can be stimulated to take risks. If research policy is to result in downstream innovation, there can be no question of keeping action on the scientific infrastructure separate from measures to promote the transfer of knowledge and its applications to the economic and social system as a whole.

But the technological imperative is still only one of the challenges presented by the new context. The shift of demand to public and private services imposes new requirements on scientific and technical research activites, which could contribute more than they are doing at present to the quality and efficiency of the social sector. Development and implementation of social innovations and technologies require special support from governments, for the simple reason that the pattern of social demand is less clear than that of market demand for goods for private consumption. Social science and technology have a role to play together here (transport, health, environment, urban development, etc) in order to improve community services, the quality of life, working conditions and the educational and cultural context.

There are few reasons for placing any conceptual limits on science and technology, but there are undoubtedly limitations imposed by

89

political, economic, social or moral factors that may delay, hamper or paralyse scientific discovery and technological innovation. 'The most intractable problems lie not in the potential of science and technology as such, but rather in the capacity of our economic systems to make satisfactory use of this potential. The success of adjustment policies will largely depend on the ability of our societies to exploit their intellectual and technological capital in responding to the social and economic challenges confronting us in the final decades of this century.'[8]

This also means that technical change is no more an end *per se* than is economic growth. 'It must find its ultimate legitimation and indispensible political support in a high degree of correspondence with the aspirations and decisions of the population of our countries.'[9] Although the public has, over the last thirty years, become accustomed to the *economic* aspects of the management of society, a great deal still has to be accomplished for it to become more aware of the implications and potentialities of *technology*.

'The demand for public participation is the legitimate expression of a more educated public in a period of profound change, which entails also changes in values, and a measure of dissatisfaction with the idea that problems can best be presented and decisions taken by the bureaucracy.' Better information and better education open the way to a more balanced perception of the technological options and issues at stake. 'Truly democratic participation is the only guarantee for our societies to overcome the resistance inevitably generated by the technical changes upon which their survival depends.'[10]

If the aim should be to restore the innovation system to health, the acceptance of a higher rate of technical change itself depends on widespread social sanction and commitment. 'This commitment will be forthcoming only if there is a satisfactory balance between the generation of new employment and the loss of old jobs and if technical change is perceived to improve the quality of life.'[11] Thus the complex of changes that defines the new economic and social context presupposes, in the science and technology field, the introduction of a new type of relationship not only between policy-makers but also between scientists, engineers, technicians and industrialists on the one hand, and trade unions, consumer organizations and representatives of the public on the other.

TECHNOLOGY AND EMPLOYMENT

I have purposely quoted all these statements, because they show how aware decision-makers have become of the practical limitations of policies that take no account of the public's resistance to technical change. Science and technology policy can no longer play to a full house in what C.P. Snow called 'the corridors of power'. Nothing is more revealing here than the debate concerning the danger of mounting unemployment as a result of advances in automation and information technology. The fact that this question was discussed at the Versailles Summit in 1982 and that it was again on the agenda for the Heads of State at their meeting in Williamsburg in 1983 shows that there is a new awareness both of what is at stake and of the sensitivity of the issue. This is not only because, as the former scientific adviser to President Carter wrote, 'science and technology were explicitly recognised as essential to the future welfare of the industrialised nations',[12] but it also reflects a realization of the growing fears for employment that are being aroused by the spread of information technology and robotics.

To allay these fears it is not sufficient to point out that both past experience and economic theory teach that it is easier to create new employment opportunities when technical progress is rapid than when it is slow, or that technology is only one of the variable determinants of structural unemployment. It is also necessary to control the process by taking measures to enable the labour force to adapt to the introduction of new technologies.

The micro-electronics revolution has resulted not only in miniaturization, but also in a reduction of the cost of rapid processing of very large quantities of information and the addition of 'intelligence' to the functioning of equipment (numerically controlled machine-tools, industrial robots, computer-assisted design and production, flexible manufacturing cells, etc). The resulting huge gains in productivity inevitably entail job displacements in the short term, if not increased unemployment (service sector examples being word-processors and other automated office equipment, electronic mail sorting, etc). In the longer term, they will lead not only to improved working conditions, but also to new jobs. These new jobs will obviously require new occupational qualifications or at any rate higher qualifications than those needed in the obsolete jobs (in the countries where micro-electronics technologies are most advanced, there is already a shortage of some categories of skilled personnel, particularly software specialists).

Nobody can say yet how long it will take for the number of jobs created to exceed the number eliminated. The answer partly depends on what the economists call the 'macro-economic compensation effects': in other words, how soon and to what extent the new technologies will lead to increased investment and from these to growth of disposable income and demand. But it also depends on the way in which manpower and education policies are able to take these foreseeable effects of technical change into account. Thus, science and technology policies necessarily extend into other fields of government action. They imply not only fresh efforts in the area of training, retraining and education, but also worker participation from the very first stages of the process of introducing and diffusing the new technologies.

In the space of a quarter of a century, therefore, blind, positivist faith has been replaced by a tempered but critical confidence in the effectiveness and virtues of the policies which direct science and technology policy. Measures taken without regard for what can be believed to be the aspiration of society – often urged by science lobbies and technocratic decision-makers – have been replaced by approaches that have more concern for social demand and resistance. Society cannot be changed by decree, any more than technical change can be imposed from on high. Industrial societies and their democratic functioning depend heavily on science and technology for their existence now and in the future, but they also depend on the way in which this issue is treated at the policy level. Taking the immediacy out of the tensions, fears and hopes, aroused by technical change – in other words, preserving compatibility, or rectifying asymmetry, between the directions imparted to technical change and the aspirations of society – is nothing more than a matter of debating and negotiating in good time the options that will decide the future.[13]

It would be unwise, however, to conclude that science and technology policy has come of age, should 'age' be equated with wisdom. The defence-linked strategic objectives of the most industrialized countries entail a large proportion of investment in military R&D. The superabundance of sophisticated weapons and the development of new weapons systems have more than a little to do with the uneasiness felt, not only by the scientific community but also by the general public, with regard to the irreversible association of science and politics. And the intense international competition in the field of technical innovation leaves inevitably little room for alternatives in the efforts to increase productivity *for its own sake*. The industrialized countries are mobilizing their efforts on the same 'fronts': electronics, computers, robotics, biotechnology, energy, materials, communication. It will not be possible for all to emerge victors from this battle to conquer new markets, and some are liable to exhaust themselves in the fray. Innovation and still more innovation remains the common goal. But would a different kind of innovation not do more to meet the deep-lying needs of a world that is more than ever divided into industrialized and underdeveloped societies, the former engaged in a race for power and gadgetry, the latter oppressed by overpopulation and undernourishment?

Never before has the process of 'creative destruction' that Schumpeter said was inherent in the dynamics of industrial capitalism been more obviously at work than in this deliberate focusing of national R&D efforts on the creation of new production functions and the development of new technical processes. The achievements of science and technology open up opportunities, but they also carry threats. How are we to know whether this stretch of the future now being determined by science and technology policies will be better mastered than before, whether it will be liberating rather than alienating, regardful of human needs rather than creative of new tensions and new dissatisfactions? The fact that these policies concern science does not necessarily mean that the paths of the future will be traced more scientifically.

SCIENCE BECOMES A COMMODITY

All these changes that science and technology policies have undergone shed some light on the matter of science as a commodity for science has indeed become a commodity. It would be impossible to understand both this status and the threats resulting from it without relating them to the policies that have been developed in this area of government intervention after 1945, as a result of the aftermath of the war, namely the ending of a world conflict without either real peace or security.

In trying to analyse the consequences for science resulting from these changes, I shall now be somewhat philosophical and question what seems to be implied in the idea of science as commodity. Is the situation we face and its related threats really so recent? Could it even have been avoided? My simple point will be to underline that it is not new and that what has happened was implied in the very nature of modern science: what we witness is but an aggravation of a process which goes back to the essence and origins of modern science.

Let me first refer to Bacon. Indeed, Bacon spoke of science as an open community of scholars, but the science he spoke of had little to do with what science meant before the sixteenth-century scientific revolution, what for instance it had meant for the Greeks. For Bacon, as well as for all philosophers who anticipated the developments of modern science – whatever were the differences of approach between him and Descartes, Pascal, Malebranche or Newton – the practice of science would be less a speculative activity than a very practical, utilitarian and efficient enterprise. Science as power, knowledge equalling power, this was the real implication of the scientific revolution. Neither Plato nor Aristotle would have thought of a science that would act upon and manipulate nature by reconciling theory and action.

Furthermore, Bacon himself anticipated that science, in being power, would lead to the developments we are discussing today. The *New Atlantis* is the dream of a science-based society in which knowledge is useful and efficient, in which researchers fulfil functions directly oriented towards utilitarian purposes. 'Salomon's House' is an island in which we find fundamental researchers as well as engineers, scientific attachés as well as managers, businessmen, diplomats, soldiers and spies.

In the book I published more than ten years' ago, *Science and Politics*, I tried to show that these developments were the beginning of our farewell to science conceived as essentially a learned and innocent activity, free from the pressures of the powers-that-be. Bacon's dreams were not fulfilled until the nineteenth century, when science became more and more associated with the industrial system and more and more confused with technology.

It was only with the Manhattan Project and the explosion of the nuclear bomb that people started to understand that science as a collective endeavour had less and less to do with knowledge and learning conceived of as a vocation whose consequences could not be harmful. And so we find scientists in academia expressing nostalgia for the 'good old days', the golden age when science was not condemned to ambivalence just as though it had not been inclined from the start to sustain itself with state, industry and military aid. While many scientists would prefer, as Don K. Price recalled, not to have to live off government largesse, to suffer from public control, and thus to practise an activity that is a means among others, they have no alternative; they are rather like the young Lady of Kent

Who said that she knew what it meant
When men asked her to dine
Gave her cocktails and wine
She knew what it meant – but she went.[14]

Science has become a tool in the heart of politics and economics, a product that can be bought, sold, traded, stolen and misused because knowledge that manipulates nature can, in its turn, be manipulated. The image of the learned scholar, the *savant* involved in research for its own sake or for his own pleasure is an image of our culture, not a reality in our societies, concerned as they are with the exploitation of research results. Science has become a technique among others and scientists fulfil functions that are today as far from those of the Greek philosophers as an abacus is from a micro computer. Bacon noted in his journal that he had no greater ambition than to serve the whole of mankind by science; he even contrasted the glory of the statesman with that of the inventor, and gave the advantage to the latter who had no problem of conscience to deal with. Three centuries later, the conquest of the atom was in turn to open up a new world where the glory of the inventor was to weigh no less heavily on the collective conscience than that of the statesman. 'Physicists have known sin,' Oppenheimer wrote, 'and that is a knowledge which they can never lose.'

The search for truth is not really the objective of modern science. Science as an instrument of power is not in search of truth, but in search of practical results. The scientist is simply one professionalized technician among many, devoted to extending knowledge whose meaning lies less in itself than in its usefulness. And since science in its relationship with industry and government has become an extremely expensive collective activity, it is no wonder that it is measured in the same way as any other commodity which has to pay off in terms of applications, profits and returns.

Where, then, are the threats? Of course the open community of scholars is under question, of course science is not simply a game, a leisure activity, a speculative endeavour whose natural location would be a university free from political pressures, far from dirty involvement in day-to-day events. The first threat is, indeed, the possibility of pressures exerted upon the substance, the content of research by those who neither know enough nor care about the process by which research can progress. The second threat is that the practice of, and therefore the support for, science will only be understood against the background of its short-term values. Scientists would prefer, like most lovers, to be cherished for what they are rather than for what they offer. But science for its own sake has no more worth than a ballet or a painting and from this point of view, it is just one of the overheads of culture, not an asset in the quest for power. The third threat is that learning and doing science can no longer be isolated in the ivory tower from the political environment and its constraints.

All these threats, indeed, have to be seriously taken into consideration. The scientific ideal, science as a vocation, knowledge as an intellectual activity within the academic framework, may suffer from being too clearly seen as a cradle of power that needs money and public legitimation. However, to my mind, none of these threats supersedes those threats which rise precisely from the very success of science when equated with power. The former category of threats concerns the life of university research, the status and functions of the researcher, the freedom of research in the open community of scholars. Important and real as it is, this category is in the end less dramatic than the latter – I mean those threats which stem from the association of science with the state, with political and military objectives, in brief from the fact that no commodity is more powerful than science, no tool, no technique more threatening for mankind.

Arthur Koestler once said that every man has always known that he is doomed as an individual, but we now know that we can disappear as a species. The real threat lies here, deep-rooted in the implications of what science has become, much more than in the limits imparted to practising science as a pure, neutral and innocent speculative activity. This is something that neither Bacon nor any philosopher of the Enlightenment could have anticipated in their Utopias for the future of science: science has become a commodity, but such a commodity is not a product like others which can be simply traded or stolen, it is also a source of potential total destruction. I should prefer to see the scientific community more concerned with what may happen as a result of what they do as technicians among other technicians than with the nostalgia for an age of scientific innocence when they had nothing to do with, nor to receive from, political, military, and industrial expectations.

That is why I think that the real threat is not that which concerns the scientific community as such, its freedom of research, its privileges as the last aristocracy in the world: namely to be paid for what one likes to do, to enjoy – as Stendhal once said of himself – the rare, exceptional privilege of turning one's passion into one's profession. Rather, the real threat is one which, through the practice of modern science, in alliance with and dependence upon the powers-that-be, can menace society and, beyond society, civilization. Is there a connection between the two categories of threats? I shall conclude by leaving open the question: are we doomed to choose between an open science in a devastated world and a threatened science in a surviving world?

REFERENCES

1. Marx K. 1970 *Capital*. English translation Moore and Aveling. Lawrence and Wishart, London, vol I chap XXV p 386.
2. Renan A. 1890 *L'avenir de la science*. Calman-Levy, Paris, p 253.
3. See in particular 1980 *Technical Change and Economic Policy*. OECD, Paris.
4. *Technical Change and Economic Policy op cit*, p 20.
5. See: Lesourne Jacques et al 1979 *Facing the Future – Mastering the probable and managing the unpredictable*. OECD, Paris; in particular Part III.
6. Faure G. 1980 Le MITI ou la politique à long terme *Revue Française de gestion*. (Special issue: 'Le Japon: mode ou modèle', 27/28, Sept/Oct) pp 51–52.
7. 1981 *Science and Technology Policy for the 1980s* Part II: Technological Innovation and the Economy. OECD, Paris, p 65.
8. *Technical Change and Economic Policy op it*, p 93.
9. *Op cit*, p 103.
10. *Op cit*, p 105.
11. *Op cit*, p 107.
12. Press Frank 1982 Rethinking Science Policy *Science*, **218**, (October 1) Washington, p 28.
13. See: Salomon J.-J. 1982 *Prométhée empêtré – La résistance au changement technique*. Pergamon, Futuribles, Paris.
14. Price Don K. 1953 (1962) *Government and Science* reissued by Oxford University Press, p 87.

SELECTED BIBLIOGRAPHY

Brooks H. 1968 *The Government of Science*. MIT Press, Cambridge, Mass

Fusfeld H.I. Haklish C.S. 1979 *Science and Technology Policy: Perspectives for the 1980s*. Annals of the New York Academy of Sciences, vol 334.

Gilpin R. 1968 *France in the Age of the Scientific State*. Princeton University Press.

Kuehn T.J. Porter A.L. editors *Science, Technology and National Policy*. Cornell University Press.

Lesourne J. others 1979 *Facing the Future – Mastering the Probable and Managing the Unpredictable*. OECD, Paris.

National Science Board 1977 *Science Indicators 1976*. NSF, Washington; 1979 *Science Indicators 1978*. NSF, Washington.

Nichols G. 1980 *Technology on Trial*. OECD, Paris.
Organization for Economic Cooperation and Development
    1971   *Science, Growth and Society – A New Perspective.*
    1980   *Technical Change and Economic Policy – Science and Technology in the New Economic and Social Context.*
    1982   *Science and Technology Policy for the 1980s.*
    1983   *Science and Technology Indicators I – Trends in Science and Technology in the OECD Area During the 1970s: Resources Devoted to R&D.*
    1966–76 *Reviews of National Science Policies*: Belgium, 1966; France, 1967; United-Kingdom/Germany, 1967; Japan, 1967; United-States, 1968; Italy, 1969; Canada, 1969; Norway, 1971; Austria, 1971; Spain, 1971; Switzerland, 1972; Iceland, 1973; Netherlands, 1973; Yugoslavia, 1976.
    1976–81 *Reviews of Social Science Policies*: France, 1976; Norway, 1976; Japan, 1977; Finland, 1981.
    1977   *Science and Technology in the People's Republic of China.*
Salomon J.-J. 1983 *Science and Politics*. MIT Press.
Salomon J.-J. 1972–4 (under the direction of) *The Research System – Comparative Survey of the Organsation and Financing of Fundamental Research*: vol I, France, Germany, United Kingdom 1972; vol II, Belgium, Norway, Netherlands, Sweden, Switzerland, 1973; vol III, Canada, United States and General Conclusions, 1974. OECD, Paris.
Salomon J.-J. 1982 *Prométhée empêtré – La résistance au changement technique*. Pergamon, Futuribles, Paris.
Spiegel-Rösing I. de Solla Price D. (editors) *Technology and Society – A Cross-Disciplinary Perspective*. Sage, London, Berkeley.
Tisdell C.A. 1981 *Science and Technology Policy – Priorities of Government*. Chapman and Hall, London, New York.
Zaleski E. others 1969 *Science Policy in the USSR*. OECD, Paris.

CHAPTER FOUR
The Interests of High-Technology Industry

Hermann G. Grimmeiss

High-technology industry normally refers to those industries which use very advanced production facilities for the manufacture of highly sophisticated products. These products can be very different, depending on which industry we are dealing with. Examples of high-technology industries are the pharmaceutical, automotive and electronic industries. Owing to their different characteristics, these industries often have different business policies. When specific questions are addressed, it is therefore not unreasonable to assume that these industries have different interests in university research. Since my experience is of the electronic industry, I would like to restrict my comments to the semiconductor industry, though it should be noted that the various electronic companies have differing standpoints which will affect their interests in academic research.

Of all high-technology industries, there are only very few which play a dominant role in the international market. The electronic industry is one of them. An unexpected miniaturization in semiconductor technology has resulted in such improved performance by electronic devices that completely new systems are now available. During the last fifteen years, the complexity of integrated circuits has increased exponentially and, consequently, the price per bit on a chip has decreased almost at the same rate. A computer of roughly the same complexity as the latest microcomputer for less than $1000, cost more than $10 000 000 in 1962. Silicon chips are now available with the equivalent of more than one million transistor functions per square centimetre for prices of the order of ten dollars. What I am trying to indicate is that even the most complex electronic circuits are now available at prices we did not dare to dream of ten or fifteen years ago. And there is more to come! The complexity of integrated circuits will probably increase by more than an order of magnitude during the next ten years, which means that chips will contain ten times more electronics for about the same price as today.

This is one of the reasons why most of the industrial and consumer products of the future will contain electronics. Although the performance of the products is often solely determined by the capacity of the electronics, the cost of the chips is in many cases negligible compared with the price of the total system. In the future, therefore, the competitiveness of a product on the open market will, in the first instance, be determined by the competitiveness of the electronics contained in the product. This is why the electronic industry holds a key position among the high-technology industries.

All industrial countries, whether in the West or in the East, are well aware of this situation. Governments in many of the these countries heavily subsidize the R&D efforts of their semiconductor industries in order to secure the competitiveness of their products. If any industry is vitally dependent on new achievements obtained in basic and applied research, it is the semiconductor industry. The research effort of companies such as IBM, Bell or NTT is well known. Previously, it looked as if the different semiconductor companies could well cope with the development of semiconductor electronics, and even lead this development. Due to the rapid progress and high investment, university research in this field was often considered as old fashioned, and to have little hope of contributing significantly to this area.

However, it was precisely this rapidly increasing cost of investment and, in some sense, also the shortage of manpower, which made the semiconductor industry rethink its situation. Increasing the complexity of integrated circuits by decreasing the line width of the pattern generation on the silicon wafers was previously a matter of scaling. However, when the line width approached one micrometre new physical problems arose which became even more serious when the line width was further reduced. New types of etching processes and lithographies had to be used, and new techniques for design and testing had to be developed. Often material problems cannot be solved by computer simulation, and therefore solutions must be found by time-consuming and tedious experimental studies. In addition, these studies often involve investigation into different branches of science, such as organic and inorganic chemistry, solid-state physics, crystallography, atomic physics etc, which present a real interdisciplinary challenge. The development of new types of lithography, such as electron beam, X-ray and ion beam lithography, demands advanced engineering which in turn involves not only optics, but also mechanics and electrical engineering. Already today's most advanced integrated circuits are too complex to be designed manually. Gate arrays and cell libraries have been developed to overcome these problems. However, it is generally agreed that new techniques will be needed when the complexity of integrated circuits further increases and more than one million transistor functions are handled on one chip. Structured design is being discussed as one of the most promising approaches, and it is interesting to note that some of the most important contributions to this new technique have emerged from university laboratories. These new design and test techniques require not only profound insight into electro-techniques, but also a sound knowledge of advanced mathematics, which is not always a strong point in an electrical engineer. It may turn out that the shortcomings of the present generation of circuit designers are not their limited knowledge of circuitry, but their limited insight into the future tools necessary for the design of ultra-large scale integrated circuits.

I could give further examples showing how complex technological progress in one of the most important, high-technololgy, industries has become, but the examples already given should demonstrate that semi-conductor electronics are quite suddenly, no longer merely a matter of electrical engineering and solid state physics, but dependent in the future on breakthroughs over a wide range of different disciplines. It is obvious that those companies which can most efficiently utilize and combine the achievements in the different scientific areas will have the best chance of becoming the market leaders. The resources needed for such an approach will probably exceed those of even the largest companies, and therefore new approaches will be eagerly sought. The individual companies will still have to perform their own product development in different areas, and for this they will have to continue their heavy investment activities. However, most of the resources for their more basic research projects are available in the many university laboratories of most industrial countries. This is the reason for the growing interest of the electronics industry in academic research.

This interest is two-fold. Firstly, the electronics industry is in need of suitably trained personnel that is sufficiently acquainted with the latest achievements in applied mathematics, physics, chemistry, mechanics, electrical engineering, computer science, etc. This does not mean that a single person needs to acquire all this knowledge, but that well-educated graduate students and experienced researchers in different disciplines must be available. Secondly, the electronics industry needs all the help it can get to make it competitive in developing circuits of still higher complexity.

Meeting these needs probably sounds easier than it is. Contemporary university education usually means that professors are well acquainted with the latest achievements in their areas, not only from a theoretical pont of view, but also experimentally. However, it should be noted that most teaching, at least in Sweden, is not performed by professors or other researchers, but by university teachers, who are not officially supposed to carry out any research at all. There is therefore a tendency, once a course is established, to continue to teach it unchanged for many years, which in some cases may mean a decade or more. We all know how fast microelectronics is developing, and outdated courses are of little worth for modern training in semiconductor electronics. Furthermore, many of our university teachers left research before the real breakthrough in microelectronics occurred, so that in fact they have never obtained any training at all in this field. Considering the complexity of the subject, it is quite understandable that they sometimes have difficulty in gaining a sufficiently deep understanding of this field. Some changes in this system may therefore be necessary if a more up to date education is to be guaranteed.

Even for courses given by experienced teachers, the quality of the course very much depends on the sophistication of the teacher's experience. In an ideal society, the accumulated knowledge of the teachers should be better than any of the experts in industry. Fifty or sixty years ago this was not an unreasonable assumption. But if the universities today are to carry out research at a level comparable with industry, then they must have at their disposal resources which are comparable with those of industry. This, however, is an unreasonable proposition given the huge investments which would be necessary if all teachers were to be included. A much better result would be obtained if electronics companies were periodically to invite university teachers to work for one or two years in the company within the field of their personal interest. For technical universities, one might consider making such training periods compulsory. Of course, such a programme would not be very useful if these teachers, when they returned to their universities, were not provided with the resources they needed in order to pass on their new experiences to their students.

So far, I have not made any distinction between courses for graduate or undergraduate students. Owing to the growing importance of microelectronics, it is my opinion that a profound insight into this subject is a necessity for all undergraduate students. The question then arises as to how much influence industry should exercise on universities and which courses should be given on which subject. I feel that decisions on the curriculum should be made by the universities only, but they would be well advised to listen carefully to any suggestions made by industry. Industry is one of the university's customers insofar as it offers a future for their products. A good curriculum, seen from the standpoint of industry, not only guarantees the future of the students, but also the competitiveness of the industry. Yet, industry is only one of the university's customers, and we should therefore not forget that universities still are, and hopefully will continue to be, the main institutions for carrying out basic research.

This brings me to the main theme of my paper: 'What interests have high-technology industries in university research?' In the present slow-down of the world economy, there is a greatly increased concern to promote research and development activities with the intention of improving industrial competitiveness. When reading scientific journals such as *Physical Review* and *Journal of Applied Physics* twenty years ago, it was quite obvious that a considerable part – if not the larger part – of the most important papers in basic and applied research was submitted by industrial research laboratories. The rapid and still accelerating development in microelectronics and the vigorous diminution in the price/performance factor in complex electronic devices, has resulted in rapidly growing numbers of new products which are frequently shortlived. Industrial laboratories were, therefore, forced to concentrate more and more on development and pilot production, and to reduce their activities in pure basic research and, to some extent, also in applied research in order to maintain their competitiveness on the open market. The introduction of an increasing number of new products forced the industry to concentrate on short-term problems to such a degree that often only small resources were left for more basic research activities.

This has a twofold effect on the relation between universities and industry. The incentive of research-orientated engineers and physicists to leave university and to join industry in order to perform basic research in areas where modern and expensive equipment is a necessity has diminished. Nowadays, the driving force for those students who leave university is the desire for higher salaries. The other effect is that universities are now considered by industry as interesting partners for solving long-term problems. It is generally accepted that an expansion of cooperation between industry and universities in the future would undoubtedly result in a deepening of the necessarily more application-orientated R&D of industry through high-quality basic research performed at universities. Since larger universities cover a broader field than individual industrial companies, universities form the natural basis of interdisciplinary studies. A further synergetic effect of university research may be achieved if their basic research is of common interest to several different companies. Since basic research normally deals with non-competitive projects, there should be no drawbacks to the results being shared by different companies.

The keen competition in industry and, in particular, between electronic companies has generated a growing reluctance to publish their achievements in R&D, and, hence, to communicate with the scientific community whether in universities or in industry. This question has even become a political issue, and scientists of certain countries have been asked by their governments to slow down communication with foreign colleagues in certain areas. Since progress in science has always been based on the international exchange of knowledge, a closer cooperation between universities and industry may also in the future guarantee such an exchange of new ideas on basic achievements.

The degree of cooperation between industry and universities seems to vary from country to country. In the USA, a number of mechanisms have been developed, which assure a strong and effective relation between the two partners. Owing to the openness that exists between universities and industry, and the willingness to support academic research financially, a high speed of transfer of scientific results into industrial products has been achieved. The situation is quite different in Sweden, where it appears that the research results of the technical universities are utilized only to a relatively small degree by Swedish industry, and that commissioned research is much less then 10 per cent of the total support given to our universities. There are many reasons and explanations for these circumstances, but unfortunately, they lie beyond the scope of this paper. Since, however, heightened interaction between universities and industry is of vital importance for the future competitiveness of our industry, I should like to give at least some of the reasons why this is so: the limited – or often non-existent – industrial experience of university staff, the diminishing economic incentive resulting from the tax system and the inflexibility of the university system, partly caused by increasing bureaucracy.

To be able to contribute satisfactorily to a rewarding relationship with industry, universities need high-quality teachers and scientists with industrial experience. Since the salaries of Swedish university teachers and professors – even after the latest raise – are very low on an international scale and in comparison with industry, clearly some difficulties must be overcome if we are really interested in a breakthrough in the interaction of industry and universities.

Assuming that these difficulties can be solved in the future, are universities really interested in devoting their activities to applied research and in developing industrial products? You will realize immediately that I have asked the wrong question. In my opinion, universities should concentrate on something other than developing industrial products. Industry is more skilled and better equipped for such tasks. Furthermore, universities should never devote all their resources to applied research, since – as I mentioned previously – industry is only one of their customers. Universities have the responsibility for independent academic research, which is one of the mainstays of cultural progress. Universities should never be deprived of their responsibility for free research, irrespective of whether or not the results contribute to our technological progress. However, it is quite reasonable to discuss how much of the research performed at universities should be dedicated to more applied projects. This percentage may vary from university to university, but for technical universities about twenty per cent is probably acceptable, provided that these projects are performed by scientists according to their personal interest. From my standpoint, this twenty per cent may even be commisioned research within basic research, as long as the proposed subjects are not defined in detail. As in the case of the curriculum, any decision on which research are performed at universities should be made by the universities or the professors involved in the project, but a useful relationship between universities and industry can only exist if universities and, in particular, scientists carefully listen to all proposals made by industry. It seems to me quite conceivable that directors of research, or technical directors of companies interested in such cooperation, could meet scientific representatives from the universities twice a year, in order to present research programmes of the highest interest to industry.

Research programmes which have been suggested by industry can be supported financially either by industry or by the relevant research council, depending on the degree of freedom the scientists in question prefer. If industry is supporting a project financially, it should still be possible to publish the results without restriction. There may, however, be exceptions. If the project is of such a character that the company initiating the project is contributing information which is crucial for one or some of their products, certain restrictions on the availability of the results would have to be made.

The question may arise as to how close cooperation between universities and industry may influence the academic environment. If financial support from industry is substantial, while the salaries and available resources at universities are insufficient, there are certain risks that at least some of the university scientists may choose more applied projects instead of performing pure research. These risks can, however, be eliminated if governments strive to generate a fair balance between the working conditions of scientists at universities and in industry. Besides, if the number of scientists at universities who wish to participate in projects financed by industry becomes unreasonably large, a natural balance would probably soon be reached due to a decrease in funding, since even industry has only limited possibilities for funding external projects.

As long as most of the results produced by university research are available to the whole scientific community, I see no harm in closer cooperation between universities and industry. Even if such cooperation were to create some competition within academic society, I can only see advantages in such an interaction, provided these working arrangements follow the rules which were discussed earlier.

Industries all over the world need modern universities with high-quality basic research, which are orientated to the international development of technology, to increase our understanding and insight into the fundamental questions of science. A rewarding cooperation can only be established if free academic research can be preserved in an environment of increasing technocracy. It will then be the responsibility of industry to use the results of university research together with their own enormous resources and manpower for the benefit of all mankind.

ACKNOWLEDGEMENTS

The author would like to thank Stig Larsson, Erik Eriksen, Bror Lundquist and Olof Sternbeck for valuable discussions. Patricia Jungstedt's and Eva Lundgren's help with the manuscript is gratefully acknowledged.

CHAPTER FIVE

The Legacy of Success: Changing Relationships in University-based Scientific Research in the United States

Dorothy S. Zinberg

Since the Second World War, the growth in the number, size, and shape of American university research departments has been nothing short of phenomenal. A once barely visible academic scientific community, charged by its success during the Second World War, burst upon the national scene. Scientific research had not only served national military needs during the war, but also had yielded the knowledge that made possible the post-war development of radar, computers, jet engines, nuclear power, and a seemingly limitless number of petrochemicals and pharmaceuticals for civilian use. The hardy scientists who had survived 'a kind of moratorium on physical research in America' when 'the money ran out' during the depression of the 1930s, lived to see and, in many instances, to share in the creation of a dazzling new world of research science.[1]

Less than a half century ago, American universities were spending an estimated $50 million a year on research in science, only 12 per cent of which was provided by the government, 'chiefly for research in agriculture'.[2] Rising sharply after the Second World War, university expenditures for research and development (R&D) increased by almost 300 per cent in constant dollars between 1953 and 1968.[3]

During the same period, the US federal government increased its expenditures for overall R&D by almost 425 per cent.[4] Since that time this growth has slowed. In the case of basic science, federal expenditures actually declined between 1968 and 1983. The decline was 3 per cent in constant dollars between 1981 and 1983.[5] However, federal funding for research and development at universities, which was about $7.4 billion in 1983, will rise substantially (about 15 per cent) in 1984, according to the proposed federal budget for 1984.[6] This is cheering news for university laboratories, which house about half of the nation's basic research and receive 70 per cent of their support from the federal government.[7]

From its origins as a small group with a 'long tradition of aloofness that separated science and government' to one that currently numbers more than one million members, of whom almost 400 000 are employed in universities, and that is inextricably bound to the federal government, the scientific community has evolved into a major force in society.[8] The President has a scientific advisor, who directs a large bureau, the Office of Science and Technology policy. The Congress has its own advisory apparatus, the Office of Technology Assessment and several science and technology committees staffed by more than a hundred scientists holding PhDs. And increasingly, state governments are creating scientific advisory bodies. The Pope, too, has an international group of scientific advisors, who in widely publicized meetings counsel him on such scientific and moral issues as nuclear-weapons technology and genetic engineering.

The American public, despite some ambivalence, continues deferring to scientists on a broad range of issues and shows greater confidence in the scientific community than in the US Supreme Court and religious and educational leaders. Only the most eminent medical doctors enjoy greater public confidence than scientists.[9]

Within two decades following the Second World War 'Little Science' as described by Derek Price – small, underfunded projects staffed by underpaid scientists fashioning their laboratory equipment from 'sealing wax and string' – had indeed become 'Big Science', a dazzling world of billion dollar budgets and many, highly visible, well-paid academic scientists whose advice is sought by governments, industry, the courts, religious organizations, and public interest groups.[10,11]

The rapid growth of science since the Second World War has altered its inner proportions (high energy physics and biomedical research, for example, now take a lion's share of the funds), its scale and its interface with society. These changes have produced new tensions. Some threaten the very form and function of science, but others, if recognized in science-policy planning, could favour rational adaptation.

CHANGES IN SCALE

In plotting the steady, exponential growth of science, Derek Price predicted in 1963 that were it to continue apace, there would be, in time, 'two scientists for every man, woman, child, and dog'.[12] There is little likelihood that such growth will occur: the federal budget for science reached its peak in 1967 and the rapid growth of research departments has slowed. Academic scientists appointed in the 1950s and 1960s, with at least another decade remaining before they retire, have effectively blocked promotion for young biologists, chemists and physicists. Seventy-six per cent of all physical science faculty held tenured positions in 1980.[13] In 1968, the corresponding figure was approximately 37 per cent.[14] If the most productive years of a scientist's life fall before the age of thirty-five, this maldistribution bodes ill for science. Only in newly emerging fields such as the computer sciences and in some branches of engineering have academic opportunities grown for younger scientists.

Today's Big Science is big not only because of its budgets and vastly expensive particle accelerator project. It is big because all of its component parts are big. More than 12 000 PhD degrees in the physical and life sciences, mathematics, and engineering are awarded each year. This figure reflects a 9 per cent drop from an all-time high, but has varied by less than one per cent for the past decade.[15]

The budget for the National Science Foundation alone is slated to grow by 17.4 per cent [to more than $1.25 billion] in 1984, although many of the beneficiaries are in the applied sciences.[16] The number of professional science magazines and journals continues to increase, and the public awareness of science – its accomplishments, its risks and errors, and its role in defence and weaponry – has become a permanent part of the scientific scene. These changes in scale have increased the public's awareness of the importance of the scientific enterprise. And because the enterprise is funded largely out of taxes, the public now demands greater accountability from scientists. Science has achieved a 'new visibility', which in turn has led to a new 'political' science.[17]

THE UNIVERSITY, GOVERNMENT, AND INDUSTRY

With the growth of science has come a change in the relationship of the university (in particular, the university research laboratory) to government and industry. As the world economy has slowed and the perceived threats to US military security increased, industry and government have looked to the university to produce the ideas, personnel and products to reverse the economic downturn and increase military security. This simple triadic relationship is built from the

109

many complexities of interdependence. Each institution has different needs, opportunities and constraints. The university, for example, with its long tradition of aloofness from government funding, shifted first from acquiescence to active search for funds, and now, to direct lobbying and less-than-genteel competition for grants and contracts.

Industry plays a smaller but increasingly significant role. Although the overall investment by industry stands at 6 to 7 per cent of total research funding, the percentage appears to be rising.[18] These funds are not distributed evenly across universities, but are clustered in two dozen or so leading research universities which receive some 11 to 15 per cent of their total funding from industry.[19]

Each of the three actors now appears intent on developing and, in some instances, exploiting every avenue of influence possible. The universities are actively soliciting financial support from government and corporations, while the public and private sectors are eagerly tapping what ideas and manpower the science faculties have to offer in order to maximize the contributions of the 'golden' scientific commodity. Indeed, their activity is now so intense that government and industry have begun to broaden their influence over the universities as they seek new ways to interact and in specific instances to shape and direct scientific enquiry.

The university is the most vulnerable member of the triad. Unique among the three institutions, universities not only produce ideas and products applicable to today's concerns, but also generate the trained personnel who will move science in new directions, and provide solutions to new problems. Thus, any change in the environment in which universities operate will eventually affect scientific inquiry. For example, when funding for basic science decreased, many young, would-be scientists in the top rank of their high schools enrolled in medical schools. When nuclear power lost its glamour, nuclear engineering lost its ability to recruit top students. Now senior nuclear engineers worry that there will be no next generation to take the lead in revitalizing the field. The results may not always be deleterious, but a changing social and political environment as well as changes in science itself, will alter the size and shape of scientific disciplines and the ways in which science is carried out.

Moreover, policies that affect university education could in time affect secondary education as well. A well-funded campaign to increase the number of American engineers and computer scientists by stressing applied science in education 'from the bottom up' at the expense of early training in the basic sciences could diminish the very pool of scientific talent that has been the hallmark of recent American science; it could also diminish the pool of talent needed to anticipate and evaluate the interaction of science and technology with society in a broader sense, that is, as history and philosophy, as well as social and political process.

GOVERNMENT AND THE UNIVERSITIES

The federal government views research universities with some ambivalence. They promise a wealth of new accomplishments in directions the government may not be able to influence or even foresee, but they also pose the continuous threat that they might disclose what the government would prefer to keep secret, or fail to achieve the empyrean goals set for them. In large measure, this ambivalence has arisen out of two conditions that have become increasingly pronounced in the past five years: 1) the fear that the US and her NATO allies are becoming vulnerable to Soviet aggression; and 2) the decline of American economic and technological preeminence in world markets. Both of these have led policy-makers into anxious dependence upon the research universities while also raising the spectre that misplaced trust in (or lax oversight of) university research could subvert the hoped-for gains and even worsen American strategic and economic vulnerability.

Accordingly, the actions of the US government towards research universities in recent years have been marked by attempts to control: 1) the type of research conducted, with emphasis on applied research over basic inquiry; and 2) the dissemination and exchange of certain information deemed vital for national economic and military security through restricting publication and foreign visas as well as cutting back international travel funds for Americans.

In its concern over national security, the US government has been attempting to exert a new degree of influence over university research. It has tried to regulate access to some laboratories funded by the Departments of Defence (DOD) or Energy (DOE), on the grounds that the broad release of information related to critical military experiments or high-technology products could compromise and diminish US-funded research efforts and thus affect national security. Recently the federal government has also begun extending its control into the more 'untouchable' areas of peer review and international scientific exchange. At the 1981 meeting of the American Association for the Advancement of Science, for example, Admiral Bobby Inman, Deputy Director of the CIA, warned that the government might require researchers in cryptography and other sensitive fields to submit their manuscripts to the National Security Association before submission to journals for publication. Clearly, this precedent would radically alter the process by which scientists alone judge whether the work of colleagues is acceptable for wide dissemination.

In international exchange, several recent examples point to the US government's increased interference in open exchange, notably the 1981 Department of State (DOS) decision to cancel the visas of a group of Chinese scientists who were to attend a conference on computer technology in California, and the March 1983 DOS action to prohibit Libyan students from taking courses at US universities that would give them access to unclassified information on aviation research or nuclear-related studies. In several instances, university scientists have reported that DOD has asked them to conduct covert surveillance of foreign visitors and limit their access to seminars and laboratories.[20]

DOS efforts to curb international exchange go beyond research involving technology that could conceivably find its way into weapons or military support systems. Late in 1981, for example, DOS informed the Massachusetts Institute of Technology (MIT) that it should not allow Mikhail Gololobov, a visiting organic chemist from Moscow, access to any MIT nutritional research facility that related to the possible production of new or improved foodstuffs, and the DOD barred him from any project in which their funds were involved. An outraged professor, Nevin S. Scrimshaw, brought the situation to public attention, but the visit was cancelled.[21]

Another concerned scientist, Stanford University President Donald Kennedy, warned that the US government is now attempting to apply 'to fundamental [scientific] research, regulations originally intended for devices or industrial processes'.[22] He, like many of his colleagues, believes that 'to apply a burdensome set of regulations to a venture that has gained such great strength through its openness will cost the nation more than it can be worth'.[23] The National Association for Foreign Student Affairs echoed this sentiment in response to the Libyan issue, and added that such attempts to restrict educational opportunities on the basis of nationality or field of interest will 'inspire distrust and suspicion' of America's educational motives and may turn many students towards the Soviet countries for the training and basic learning they desire.[24]

In addition, the government is exerting a degree of control over the direction of scientific research in the name of enhancing military security. The most visible example of this phenomenon is the new emphasis on 'star wars' technology. The Reagan administration recently proposed a substantial 10.3 per cent increase in funding for the DOD's basic research programmes, thereby providing a clear signal to grant seekers.

In June 1983, Reagan's chief science advisor George A. Keyworth II said, 'The research community has an important role to play in this country's future,' but only if it will 'come to grips with the realities of the 1980s'.[25] These 'realities', undoubtedly include producing ideas and hardware to support US strategic and military aims and accepting the tighter controls on the publication and communication of their findings to the scientific community and the public. The 1984 Defence Bill recommends that $30 million be spent in updating laboratory equipment but the funds are to be given only to laboratories carrying out defence-related research.

The US government has acted on the process of scientific enquiry by invoking threats to economic security and technological preeminence. In one instance, the government has fostered a 'crisis mentality' with respect to foreign technological competition, most notably the perceived threat of the Japanese automotive and home-electronics industries. This reactive approach has led to a disproportionate emphasis in engineering schools and industry on the products of advanced technology – the fuel-efficient 'roomy' car or the personal stereo system – rather than on the 'capacity for technological innovation – the ability to continuously discover and refine . . . frontier technologies'. As the National Academy of Sciences (NAS) panel on Advanced Technology, Competition and the Industrialized Allies reported in 1983, this overemphasis on tangible results deprives American science of the 'integrated learning' necessary for long-term technological superiority under rapidly changing conditions and markets.[26] The constant comparisons to other countries also only reinforce the mistrust of foreign nationals at US universities that is encouraged by current US strategic policy.

Some good may come from these trends. The government's efforts to restrict and direct scientific enquiry have by their clumsiness alerted the scientific community in a way that more insidious control might not. And perhaps, the American people need to have their faith in technology fortified. As an NAS panel said recently, scientists must help 'maintain the historical belief of Americans in their country's ability to adapt to her circumstances, to compete, to continually expand technological frontiers'.[27]

INDUSTRY AND THE UNIVERSITIES

Dramatic changes in the 'triangle' are occurring in the industry –university relationship. Big business has discovered the vast pool of valuable information residing within university science departments and the scientists have rediscovered industry. In many instances, the lines between scientist and businessperson are beginning to blur as academic researchers, once largely confined to the role of consultant are becoming industrial researchers within their university laboratories and corporation presidents in the business world. Ten years ago, a biochemistry department might have imagined losing its most gifted scientists to another university or to a government-funded laboratory, but who would have imagined that they would soon have the opportunity to become chief executive officers of their own corporations? The model has existed for engineers in academia who have traditionally had connections with industry or owned their own corporations. Even the field of economics boasts an academic entrepeneur who recently sold his company for $110 million; and there have been scattered examples in other fields. Now, however, a major segment of molecular biology seems to have become an applied science.

The ability to bring in grant money has been a requisite in leading universities; but the sums were often modest or modest in comparison with what is currently occurring. Now that single corporate grants exceed $70 million, once-obscure scientists suddenly have the potential to improve the prestige and financial health of a university to an unprecedented extent. A New York State University president, blocked in his efforts to save $10–15 million for a laboratory for a 'star' researcher, lamented that he was in danger of losing him to competing universities. A 'star' attracts 'top students, talented graduate students and Federal grants'. [He] also attracts private money because 'many companies would like to have their name associated [with him]'.[28]

Even more complicated is the state of university–industry relations where the research is at the cutting edge of manufacturing technology. For example, the basic research for the next step in semiconductors is in the field of very large scale integration. Two major universities have each invested $20 million in pilot projects; industry has also contributed funding in order to have access to the research and to the trained students whom they would like to employ in the future. Industry representatives argue that it is necessary to limit access to these university laboratories, because the science being carried out in these universities will determine whether or not US electronic-computer industries can survive. The security issues – economic and military – they insist, are paramount; universities cannot pursue work which becomes either directly competitive with industry (some of the research yields marketable products) or poses a threat to it.

A new building at Stanford University has become a symbol of the tightening relationship between academia and industry. Nineteen corporations will have 'facilitated access' to the research results, will see the work in advance, and will be able to 'station full-time representatives on the campus'.[29] The industries are paying for the building, but DOD is supporting the research at the level of $12 million annually. The ensuing debate about academic freedom is intense. On one side, the supporters argue that 'getting support from industry enhances academic freedom because it broadens our base of support'. On the other side, critics argue that 'there is a danger that researchers will create relationships that are likely to influence what they study and what they do not study. It is a threat to the autonomy of the university'.[30] An experienced scientist, former government official, and now industrialist commenting on the heated debate stated:

> Universities have not faced up to realities. The very nature of university research is changing and these changes must be discussed in a meaningful, non-flag-waving way between industry and academia.[31]

Clearly the discussions must also take place within the academic community.

As the US National Science Board said in its fourteenth annual report to Congress late in 1982, 'We may be at the threshold of a permanent new state of corporate–academic research relationships'.[32] Despite the recessionary economic conditions of 1980–81, the report noted, the total corporate contribution to academic research and development totalled over $400 million, or some 7 per cent of all R&D expenditures, and shows signs of growing even faster in the near future. Even where industry is not funding university research directly, it has achieved the same end by hiring faculty members as part-time consultants. The practice is time-honoured; only the fields change. A *New York Times* article referred to the 'rapid and comprehensive entanglement with industry of late', and speculated albeit inaccurately that not a single top researcher in molecular biology remains without some commercial tie.[33]

Two concerns have prompted the unprecedented vigour of industry's interest in university research. First, as the 1982 National Science Board report pointed out, the production and innovation processes in many industries have now evolved to such complexity that the understanding of fundamental physical and biological phenomena is necessary to sustain these processes.[34] Consequently, corporations need access to highly skilled basic researchers (as opposed to the trial–and–error inventors of previous eras). And they need interdisciplinary expertise that no single industrial laboratory is likely to encompass. Secondly, as the NAS panel noted, today's advanced technologies are characterized by rapid obsolescence, which compels companies to organize flexible research teams that can repeatedly ensure a quick return on the company's investment.[35]

What are the special pitfalls of this increased industry–university cooperation? Observers are ascribing almost apocalyptic consequences to some aspects of these ties (as they have to the new products of science themselves) even as they acknowledge the potential benefit of the new arrangements. As the federal government wavers in its support of basic university research leaving researchers uncertain about sustained funding, the corporations become increasingly attractive to scientists and harried university administrators burdened by rising costs and deficits.

Consequently, universities may increasingly find themselves working on mission-oriented applied research for companies to the detriment of basic research. As the philosopher, Arthur Caplan, has said, '[in an industrial setting] research is pushed in the direction of productivity – not towards what will lead to the greatest scientific knowledge or what will do the most good for those who have the greatest need'.[36] Caplan is concerned that since 'companies rarely encourage fishing expeditions', the more industry funds research, the less US universities will nurture 'the capacity to innovate'.[37]

Individual science departments may become skewed towards the subfields that promise practical gains. It is easy to envision biology departments concentrating less on animal behaviour or evolutionary theory and more on molecular biochemistry and applied genetics. Universities might also be forced to base some of their hiring decisions on the perceived ability of each candidate to bring in corporate funds.

Another serious concern about corporate funding is that investigators usually lose the opportunity to submit their proposals to peer review. The director of MIT's Whitehead Institute, David Baltimore (see below) agrees that when researchers move outside the government-sponsored research system, they are unable to establish a track record of review by recognized experts in their field. This condition could be detrimental to younger scientists who become part of the social network in their fields through this process.[38]

Critics of the new industry–university partnership worry that university researchers may not only spend less time on teaching and basic research, but that they may also lose objectivity and credibility as impartial commentators on both their specific area of expertise and on national issues in general. In recent years a scientist who owned several hundred-thousand shares of a company that stood to benefit financially from relaxed National Institutes of Health guidelines on recombinant DNA research testified in favour of abolishing such guidelines on scientific ground. The American public has turned repeatedly to scientists for reassurances or warnings about the implications of new technologies as diverse as mutant organisms and nuclear missiles. Now, according to Tufts University Professor Sheldon Krimsky, fewer unaffiliated scientists are free to question publicly where and how fast particular risk-prone technologies are moving.[39]

Finally, increased coporate-sponsored research might lead to the institutionalization of secrecy in universities, requiring researchers in the same laboratory to hide their work from each other because the work is important to two different firms. The president of Stanford University Hospital was quoted as saying, 'The motive force of industry is profit . . . and the mode is secrecy and proprietary control of information', whereas the university has traditionally depended on open exchange of ideas.[40]

Two of the most prominent industry–university arrangements of the past several years demonstrate that concern about these problems is proper. They also demonstrate the flexibility of the corporate sponsor in overcoming many of them and the determination of the university to attempt to maintain intellectual integrity and academic freedom. As the two examples discussed below are among the most careful and balanced agreements forged to date, it is unclear whether the merits or demerits of these cases will remain at the forefront of the debate among academics for very long, or whether they will be supplanted by more egregious examples of industrial control of university research.

Hoechst AG/Massachusetts General Hospital.

In 1981, one of the world's largest chemical companies, Hoechst AG of West Germany, paid Massachusetts General Hospital (MGH), a Harvard University teaching hospital, nearly $70 million to set up and staff a department of molecular biology. The substantive *quid pro quos* Hoechst would gain from the arrangement were significant: a guarantee that if the laboratory produced one or more ideas or inventions that could be developed into a marketable product or technique, Hoechst would have the exclusive option to develop and sublicense it; the right to restrict the laboratory's research to molecular biology; the right to send up to four Hoechst trainees to the laboratory at any one time to work and to observe the US scientists; and the right to scan all articles before they were submitted for publication in order to determine whether they contained potential marketable products. Hoechst did not insist on the right to veto any candidate for the laboratory who was recommended by a Massachusetts General Hospital search committee and approved by the Harvard Medical School. The Hospital also retains the right to apply for National Institutes of Health (NIH) or other industrial funding if Hoechst should decline to finance additional research beyond the financial assurances in the agreement.

Whitehead Institute.

In November 1981, the faculty of the Massachusetts Institute of Technology (MIT) voted (reportedly by an 8:1 margin) to accept a grant-plus-endowment offer of $120 million from Edwin Whitehead, founder of Technicon Corporation, a laboratory instrumentation firm. Whitehead's gift was used to establish the Whitehead Institute for Biomedical Research separate from the biology department because he did not wish simply to turn over the money to MIT directly. The benefactor insisted that his motives were 'purely philanthropic'. According to Whitehead, a direct gift would have diluted its impact because he believes that MIT's primary responsibility is education rather than research. Some tie with MIT was also necessary, Whitehead felt, to help supplement the institute's twenty scientists with a 'critical mass' of ancillary scientists. A separate institute allowed his family broad representation on its board; a direct gift to MIT would not.

The institute is likely to be less concerned than a separate profit-making corporation would be with the short-term financial fruits of the scientists' research. Nevertheless, some MIT faculty are concerned that the institute will make the initial choice of areas of research, and may thus have a significant long-term effect on the make-up of the biology faculty at MIT.[41] Others are concerned that faculty will increasingly be selected for their research contributions rather than for their teaching and research ability, and that the dual allegiance and increased status of the professors with dual appointments could lead to tensions within the faculty.

Other joint ventures.

The univesities' concern about these meticulously crafted and un-usually favourable agreements are overshadowed by the problems some less perspicacious agreements have caused. Many companies with high expectations of profit have engaged the help of research universities, and not all of the firms have respected traditional univer-sity prerogatives. Vanderbilt University, for example, recently re-fused to accept a corporate research contract when the sponsor tried to limit severely the findings that the university could publish.

As with the case of government-policy shifts towards universities, the rise of corporation–university ties has produced a number of constructive responses. Universities have begun to promulgate thoughtful guidelines on how professors should conduct themselves when trying to balance school and corporate sponsorship. For ex-ample, Harvard's guidelines, published in May 1983, prohibit faculty from conducting secret research, require researchers to disclose the type and extent of their commitments and remunerations, and seek to ensure the development of inventions 'fully and rapidly in the public interest' by putting the burden of proof on the corporate sponsor to show why an exclusive licence is necessary (and allowing Harvard to 'march in' if it feels an exclusive licensee is not 'commercializing the discovery satisfactorily'). Many of the 'new' problems of profit-direc-ted research and institutional secrecy have always existed in some degree. The academic 'publish or perish' dogma encourages res-earchers to hoard some of their findings until they have been published. In addition, senior scientists need to publish 'first' in order to secure research funds and the other rewards of scientific excellence – Nobel prizes, presidential awards, prestigious university chairs, and public visibility. The new guidelines are needed to guarantee that the delicate balance between secrecy and openness which has been largely maintained in academic science will not be tipped towards secrecy as corporate funding increases.

GOVERNMENT AND INDUSTRIAL RELATIONSHIP

The US federal government has encouraged the rapid rise in corporate support of university research, because it frees large sums of federal revenue for use elsewhere without arousing the ire of special interest groups concerned that the universities' resources should not be depleted. The process is also integral to the Republican administration's commitment to increase the participation of the private sector and decrease that of the government. In addition, both government and industry have a vested interest in preferring practical and applied research to more esoteric fundamental enquiry. This is a kind of 'what's good for General Electric is good for America' revisited, and it is especially cogent in the light of the declining productivity of some traditional US industries.

Since 1980, private industry has overtaken federal government spending for all research and development reversing the twenty- year dominant role of the government. In 1983, according to the estimates of *Science Indicators – 1982*, industry spent \$44.3 billion and the federal government spent \$39.6 billion.[42] Nevertheless, the federal government remained the larger funder of basic research, \$7 billion compared with industry's expenditure of almost \$2 billion.

In this climate of shared purpose, government and industry reinforce each other's efforts in three ways:

1. Government can selectively 'deregulate' in those areas, both substantive and procedural, where a freer hand will provide the kinds of arrangements for research that both industry and government find appealing. For example, in December 1982, the National Institutes of Health approved a Harvard Medical School research protocol involving the transfer into *E. coli* of the *Corynebacterium dipththeriae* gene that produces the deadly diptheria toxin. Allowing the creation of a common intestinal bacterium endowed with the ability to secrete diptheria toxin would never have been entertained five years ago, when speculation about 'Andromeda Strain' mistakes of genetic engineering was widespread. Because the Harvard experiment is intended eventually to produce a hybrid protein that will selectively destroy human cancer cells, the government decided that 'bending' the biosafety rules was appropriate in this case. Government may also intervene to ease procedural problems that are impeding industry –university relationships: the 1983 National Academy of Sciences panel recommended that the US reexamine its antitrust policy because it might be interfering with beneficial joint ventures.[43]

2. Since many universities remain dependent on both government and corporate funding, government actions can increase the influence of domestic firms on American schools. For instance, the Investigations and Oversight Subcommittee of the House Science and Technology Committee held hearings in 1981 and 1982 into the

Hoechst–Massachusetts General Hospital agreement, on the premise (as subcommittee chairman Albert Gore, Jr, noted) that a foreign company might be 'skimming the cream produced by six decades of taxpayer-funded work' if equipment or manpower funded by the National Institutes of Health contributed to the creation of an invention over which Hoechst gained exclusive rights. Although the Gore subcommittee found no improprieties in the arrangement, the government's message to the American university seemed clear – enter into relationships with US companies instead of foreign firms and there will be no need for intense scrutiny over the strict demarcation between government and non-government work.

3. Stanford's president Donald Kennedy has noted that recent government efforts to restrict the openness of the international scientific enterprise have 'coincide[d] with efforts of private sponsors to expand secrecy for proprietary reasons'.[44] According to Kennedy, part of the effect of government's successfully restricting open exchange will be to make it more difficult for universities to resist the growing demands for secrecy by the private sector.

INTERNAL PRESSURES

As the basic research enterprise in the American university struggles to find an equilibrium in the face of changes in scale and external threats, new pressures from the universities themselves complicate the adjustment. For example, if basic research is defined as that which promises to lead to high technology which in turn will lead to new industries and an improved competitive position for the US in world markets, then basic science is more likely to be perceived by funders as that which is closer to applied science, namely fields such as materials research; computer, electrical, and systems engineering; and mechanical engineering and applied mechanics. The last two fields have had their NSF-funded grants increased by nearly 30 per cent in the 1984 budget.[45] The less seemingly applicable sciences, zoology, animal behaviour, plant physiology, or entymology – all fields which have made significant contributions to knowledge – could be permanently shrunk with little likelihood of attracting gifted young scientists to their ranks.

In other words, the changes in emphasis in one form of science over another within the university could become institutionalized thereby seriously eroding the vitality of several important disciplines.

The research community suffers from several forms of internal tension. One is the rift between factions within the academic community and another is more personal, the psychological pressures confronting the individual researcher. These complications arise both from new demands placed upon scientists and from the magnitude of the rewards available to researchers who become successful, as defined by tenured professorships, awards and prizes, and large financial gains.

RIFTS WITHIN THE ACADEMIC COMMUNITY

The new 'star' status of science faculty, created partially by the tremendous financial rewards industry can bestow and partially by nouveaux-riches universities in their search for big names, has stimulated a novel form of competition among universities. It has till now been reserved for outstanding athletes. The Ivy League universities of the northeast are inured to occasionally losing a tenured professor to a multinational corporation, but within the space of one year, one university lost not only a biology professor to a multinational company (his own), but also a Nobel Prize winner in theoretical physics lured away by an unprecedented salary offer from a burgeoning southwestern university. (The Ivy League colleague with whom he shared the prize is working part time at a rival university in the same state.) Both men are reputed to be earning more than $250 000 per year or, as the rumour goes, just under the salary paid to their universities football coaches.

Market forces have always been a part of the shifts among posts in US academia, but now the scale has dramatically altered the significance. These changes have created rifts within the faculties of each institution; the humanities and social science faculties often feel that they are being neglected, while those scientists who are working in noncommercial fields or who have been less successful in acquiring grants, also feel that they are losing influence in their departments and in their universities.

The presence of more than 300 000 foreign students is beginning to increase tensions in universities. The majority of scientists value free exchange of information as an absolute good and have been articulate and audible in defending this position. There is a minority whose grants and in some instances personal ideology have led them to be more concerned about the loss of industrial-related information or technological information relevant to military security. This is a new form of tension which, stimulated by government and industry anxiety is likely to grow.[46]

PSYCHOLOGICAL PRESSURES

The competition for academic positions as well as for funds is intense, and the psychological pressure for a number of scientists has proven to be unbearable. One of the manifestations has been a rise in the number of cases of academic fraud. In a recent book that has stirred up as much controversy as it examined, the authors William Broad and Nicholas Wade claim that cases of academic fraud have begun to 'crop up so often they can't be dealt with as isolated events'.[47] Among the factors contributing to this alleged rise in fraud, which includes both falsification of data and plagiarism of others' work, Broad and Wade cite the emphasis of certain laboratories on the sheer quantity of academic papers that are produced in order to win the competition for funding. In addition, the authors believe that the lack of peer review which results when research arrangements bypass traditional review processes has also led to an increased incidence of fraud. They quote, for instance, a researcher who when found to have fabricated data on new drugs he claimed to have synthesized, offered the following excuse for his conduct: 'I had to earn the money for research, or die.' Even top scientists, as John Ziman of London's Imperial College of Science and Technology explains, can sometimes have their 'notorious tendency towards self-deception' reinforced by glamour and money, and become 'tempted to skimp their duties as patrons and colleagues in research'.[48]

Although the question if whether fraud is actually increasing is still open – Ziman contends that it is merely more visible now as demands for public accountability increase – the temptation to commit fraud demonstrates the kinds of risks to basic scientific enquiry that increase as science grows and the pressures from within and without grow with it.

CONCLUSIONS

The conference on *Science as a Commodity* that occasioned the writing of this chapter posed the question: 'Is the spirit of scientific enquiry at risk at American research universities?' the present answer – in the affirmative – explains this risk in the light of the rapid growth and changing nature of science, and of the changing relationships between science and industry, governments foreign and domestic, and the American public.

Universities have not been passive in the development of these new and controversial relationships. They have lobbied for government funds and initiated some of the most visible arrangements with industry. As federal support for basic research declined over a decade, universities obtained government and industry funding for work in such new areas as applied technology.

From these new relationships much good has come – striking scientific achievements and hundreds of PhDs well trained on government grants. And the threats to free inquiry have provoked a healthy new measure of vigilance.

The major research universities have issued guidelines on military (or classified) and industrial contract research. The National Academy of Sciences (NAS) and the National Science Foundation have published reports alerting the scientific community and the public about threats to the freedom of scientific inquiry. The National Academy of Sciences has established a permanent, independent group, the Government-University-Industry Research Roundtable, to analyse the complex institutional issues and to explore alternative approaches. And the dangers of secrecy have been examined at length. More recently, the American Association for the Advancement of Science has initiated a project on *Secrecy and Openness in Science* which will bring together a group of academic scientists, philosophers, lawyers, historians of science, and science policy experts to research these problems. Their findings will be published in 1985.

Even internal problems of the practice of science, long protected from publicity by the scientific professions, have been subjected to open discussion. One example, the apparent increase in cases of scientific fraud, reflects the issues discussed in this paper. Is the increase simply a result of changes in scale? (The larger the enterprise, the greater the number of scoundrels.) Or are there more revelations, as a result of the larger interface between the scientific enterprise and the public? The weekly journal *Science* has been relentless in tracking down each case of fraud, and newspapers have been quick to report that all is not well in science.

Throughout the history of science, chicanery and mediocrity have coexisted with heroes, heroines, and genius. As long as the scale of science was modest and its products valued but within limits, the myth of science as disinterested, selfless, communal and universal could be maintained.[49] But the combination of the changing scale of scientific research and the psychological overinvestment by government and industry in what they perceive to be the wellspring of America's very existence, has created the conditions for a destablilizing of university science. Though powerful in resources and accomplishments, the university remains a vulnerable institution whose long-fought-for freedoms still require vigilant protection. In that regard, the university is unchanged.

ACKNOWLEDGEMENTS

I wish to thank Adam Finkel whose research, editing and writing skills contributed significantly to the early stages of the project, and Professor Harvey Brooks whose careful reading of the paper provided valuable insights and relevant data. In addition I want to thank Mary Ann Wells and Diane Asay, for their patience and professional word-processing skills which made possible the many iterations of the paper.

REFERENCES

1. Quoted in Kevles Daniel J. 1978 *The Physicists: The History of a Scientific Community in Modern America*. Alfred A. Knopf, New York.
2. Quoted in Greenberg Daniel 1967 *The Politics of Pure Science*. The New American Library, from *Research – A National Resource: I. Relation of the Federal Government to Research*. National Resources Committee, December 1938, p 65.
3. Smith Bruce L.R., Karlesky Joseph J. 1977 *The State of Academic Science: The Universities in the Nation's Research Effort*. Change Magazine Press, New York, p 18.
4. *Ibid*, p 16.
5. Shapley Willis H., Teich Albert H., Weinberg Jill P. 1983 *AAAS Report VIII: Research and Development, FY 1984*. Washington, American Association for the Advancement of Science, Washington DC, p 11.
6. National Science Foundation 1983 *AAAS Report, Science Resource Studies: HIGHLIGHTS*. NSF Washington DC (July 22), p 4.
7. *Ibid*.
8. Greenberg *op cit*, p 000

9. 1982 *General Social Surveys: 1972–1982*. National Opinion Research Centre, University of Chicago, pp 111–114.
10. de Solla Price Derek J. 1963 *Little Science, Big Science*. Columbia University Press, New York and London.
11. Weinberg Alvin 1967 *Reflections on Big Science*. Pergamon Press, New York.
12. Price, *op cit*, p 17.
13. 1981 *Science Indicators 1980*. Report of the National Science Board, Washington DC, p 139.
14. *Science Indicators 1982*. p 138.
15. National Research Council 1983 *Summary Report 1982: Doctorate Recipients from United States Universities*. National Academy Press, Washington DC.
16. Shapley, others *op cit*, p 59.
17. For a discussion of the 'new visibility' see Holton Gerald Limits of Scientific Inquiry *Daedalus* (Spring 1978).
18. Industry Support of Academic Research Growing *Chemical and Engineering News* 21 Feb. 1983, p 18.
19. Casey Susan Facing Cuts in Federal Grants, Big Schools Try to Get Research Work from Business *Wall Street Journal* 9 Feb. 1982, p 27.
20. Personal communications and Gelbspan Ross When Scientists Get Aid from U.S. *Boston Globe 23 Jan. 1984, p 1*.
21. *Science and Government Report,* **12**: No 17, October 15 1982, p4.
22. *Science* **216** (4544) 23 April 1982.
23. *loc cit*.
24. *NAFSA Newsletter* April/May 1983, 126. Letter from NAFSA Executive Vice-President John F. Reichard to US Secretary of State, George P. Shultz. The Libyan student incident is more troublesome than most. They had violated the stipulations on their visas which had been granted with the condition that they should not enroll in nuclear engineering and aeronautical courses.
25. *Science* **220** (4602) 10 June 1983, p 1123.
26. From 1983 *International Competition in Advanced Technology: Decisions for America* 78 pp, reported in the *National Academy of Sciences (NAS) News Report,* (ISSN 0027–8432) **33**, (4) April 1983, pp 3–5.
27. *Ibid*.
28. Weinrip Michael Bureaucracy Slows Stony Brook Effort to Keep Top Professor *The New York Times* 13 Jan. 1984, p B2.
29. Reinhold Robert Stanford and Industry Forge New Research Link *The New York Times* 10 Feb. 1984, p A22.
30. *Ibid*.
31. Remarks by Norman P. Neureiter of Texas Instruments delivered at a National Academy of Sciences meeting on *International Cooperation in Science and Technology*, 29 Sept. 1983.
32. *University–Industry Research Relationships: Myths, Realities and Potentials*. National Science Foundation, US Government Printing Office.
33. Bouton Katherine Academic Research and Big Business: A Delicate Balance *The New York Times Magazine*, 11 Sept. 1983, p 63.
34. 1982 *University–Industry Research Relationships: Myths, Realities, and Potentials*. Report of the National Science Board, p 16.
35. *NAS News Report* **33** (4) p 3.

36. Quoted in Bouton *op cit*, p 121.
37. *loc cit*.
38. *Ibid*, p 152.
39. *Ibid*, p 153.
40. *Ibid*, p 63.
41. Norman Colin 1981 MIT Agonizes Over Links with Research Unit *Science* **214** (23 October) p 417.
42. *Science Indicators 1982, op cit*.
43. Reported in *NAS News Report* **33** (April 1983) p 6.
44. *Science* **216** (4544), 23 April 1982.
45. AAAS Report, *op cit*, p 1.
46. For an extended discussion of the issues related to the education of foreign science and engineering students in the US, see Zinberg Dorothy S. Sending Ideas Abroad: The Education of Foreign Students in Science and Engineering, in International Council for Science Policy Studies' volume in honour of Derek de Solla Price and Radovan Richta (forthcoming).
47. Broad William Wade Nicholas 1983 *Betrayers of the Truth: Fraud and Deceit in the Halls of Science*. Simon and Schuster, New York.
48. Ziman John Review of Betrayers of the Truth *Times Literary Supplement*, September 9 1983 p 955.
49. Robert K. Merton identified the norms of science as univeralism, communism, disinterestedness, and organized scepticism in 1942. Merton Robert K. 1973 The Normative Structure of Science *The Sociology of Science: Theoretical and Empirical Investigations*. University of Chicago Press.

PART THREE

Individuals and Institutions Re-examined

CHAPTER SIX

Science and Society – A Changing Relationship

©Emma Rothschild, 1983

Is science more of a commodity now than it was ten or fifteen years ago? The pressures discussed today may rather be, from an economic persepective, a predictable consequence of the principal belief of the last generation, linking science, economic growth and the national interest. One of the central articles of faith of the post-war period, shared in different countries and by people of different political views, suggests that science is, in some sense, the main source of economic growth. This article of faith has been interpreted in different ways, politically and economically, and it has been taken as having very different sorts of practical implications.

There is what could be called the political version of the central propopsition that science is the source of economic growth. The most vivid illustration came in the famous study by the American scientist Vannevar Bush, published immediately after the Second World War. Bush had been to a great extent responsible for the mobilization of American science during the war. His study was called *Science, the Endless Frontier*, and in it he applied the principles of wartime to the problems of peace. 'New products, new industries and more jobs require continuous additions to knowledge of the laws of nature and the application of that knowledge to practical purposes.'[1] This idea of science as a new and endless frontier immediately became very popular. But it is worth noting that the metaphor of science as an endless frontier was several decades older than Vannevar Bush. Frederick Jackson Turner, the celebrated American historian who observed the closing of the western frontier and of the territorial expansion of the United States in 1890, actually turned to this theme in his own later writings. In 1914, for example, he wrote, 'In place of old frontiers of wilderness there are new frontiers of unwon fields of science.'[2] And of course this was an optimistic view. He believed that the frontier of science would replace the frontier of uncultivated land and make all sorts of future progress possible: 'the test tube and the microscope are needed rather than axe and rifle in this new ideal of conquest.'[3]

These political declarations of faith have a close economic counterpart and there are many examples that I could cite. I will mention only one which occurs in the work of the great economist Simon Kuznets, who is particularly relevant because as the leading historian of modern economic growth his views are relatively uncontroversial among economists. For Kuznets, '[a] high rate of growth in the stock of useful knowledge and of science itself [is] the major permissive factor in modern economic growth.'[4] Increases in the stock of useful knowledge and of science are 'a necessary if not sufficient condition for further high rates of growth of per capita product and productivity.'[5] It is easy to see, even in these most general and most classic examples of the principle of science and economic growth, that science is already considered as a commodity-like thing. In the political version the metaphor is of science as being like land, and land, of course, is something that can be bought, that can be spoiled, and so forth. In the economic version, the metaphor is of science as the equivalent of capital, that is to say, the equivalent of machines and other forms of equipment which can be bought, sold, and stocked. The stock of useful knowledge, the stock of science, in Kuznets's phrases.

CHANGING PERCEPTIONS OF R&D IN ECONOMIC GROWTH

I suggest that one could well have predicted the subsequent pressures on science on the basis of even these rather early and general explications of the relationship between science and economic growth. When one looks further at the practical applications of the perceived relationship, the pressures become even clearer. Take as a first example the economists' own views. Numerous theories of economic growth in the 1950s and 1960s assigned pride of place to something like technical change or a residual factor. Such a residual might be identified with the advance of knowledge, presumably captured by expenditure on research and development. It might be identified with advances in education — sometimes, in different theories, equivalent to that very revealing quantity, 'human capital'. These theories are now considered unsatisfactory in different ways. In particular, attempts to demonstrate a satisfactory causal connection between expenditure on research and development, innovation, productivity growth and economic growth itself, have been unedifying. The associated body of economic theory was nonetheless one of the principal justifications for the massive expansion both of public support for science in the 1950s and 1960s and of science and technology policies throughout the OECD countries and in many developing countries as well. Public support for university and other research programmes was justified on the basis of the national interest and in particular the national interest in future economic growth.

Of course, not every practical interpretation of the basic relationship between science and economic growth was optimistic. As a second interpretation, consider the concerns which have gone right through the modern period about automation and the possible adverse consequences of technical progress for employment. These preoccupations were rather intense in the 1950s, but they date back further than that. The mathematician Norbert Wiener, in his autobiography, describes what he calls the moral problems of the scientist, and he mentions two such problems which seemed of overwhelming importance to him in the 1940s. One was (and this is predictable enough) the atomic bomb, and in his case, the possible application of his mathematical work to nuclear weapons development. The other problem, and this is much more surprising, was the possible application of his cybernetic theories to the automatic factory and the consequences thereof for future unemployment.[6]

A third interpretation of the relationship between science, technology and economic growth was also pessimistic. This was current in the late 1960s and early to mid-1970s and was associated with the view that technology and the sciences on which it depended were the major cause of environmental pollution, the depletion of natural resources, the deterioration of the quality of life. These very serious problems were thought to be a direct consequence of the virtuous circle leading from science to economic growth and the remedy suggested was to interrupt the circle, whether by having less science, less technology, or less growth.

The fourth interpretation, chronologically speaking, and the most recent one, is once again optimistic. But it is this fourth interpretation that is, I think, at the source of the problems which are discussed in this volume. It suggests that technological development and increased innovation can provide some sort of solution to economic problems even in the short or medium term. That is to say, in the present economic crisis, increased support for science and technology provides one way out, one source of new jobs, and above all of increased international competitiveness. In particular, such a view suggests, governments should do everything they can to support their own high-technology or science-intensive industries and the research that lies behind those industries. This view, in some form or other, has been espoused by many different governments in different countries, certainly by the Mitterand government in France, in dramatic form, and in the United States, where this tendency within the Democratic party is known after the celebrated video company as the 'Atari Democrats'. Some fifteen countries, at a rough guess, have in the past few years determined to support high-technology industries and in particular those based on biotechnology and electronics.

What are the consequences of these interpretations of the relationship between science and the economy for scientific and technical institutions and the scientific community? I must start by following others in emphasizing that there is indeed no idyllic lost scientific community. I was curious, in thinking about science as a commodity, to look back to what Bacon himself had written, how his vision stands up to what is described several times in this volume as the Baconian scientific ideal. Bacon wrote: 'We maintain a trade, not for gold, silver or jewels; not for silks, not for spices, nor any other community of matter; but only for God's first creature, which was Light.' However, what he described as the trade in light was hardly the activity of a pure scientific community. His wise men, like any rapacious industrial powers of the late twentieth century, sought knowledge 'of the sciences, arts, manufactures and inventions of all the world; and withal, to bring unto us books, instruments and patterns in every kind.' They even had their own spies engaged in covert commercial activities. Twelve fellows 'that sail into foreign countries under the names of other nations (for our own we conceal) who bring us the books and abstracts and patterns of experiments of all other parts. These we call Merchants of Light.'[7]

Much has been made of the internationalism of the scientific community between Bacon and the Second World War. But, as Jean-Jacques Salomon has pointed out, the fact that Humphry Davy was able to speak at the Institut de France during the Napoleonic Wars was more evidence that the governments of the time were scarcely aware of the services that science could render to them.[8] A useful scientist at the time, or rather a useful amateur engineer, was, for example, the American inventor Robert Fulton, who produced one version of a submarine for the French government during the Napoleonic War, and then moved to England and produced a better version for the

British government. He had a great deal more difficulty in circulating and ended up travelling under a number of elaborate disguises, false names, and so forth.[9] People who were really useful and who also came from social classes other than those that produced scientists and inventors were treated in a quite different way. Think of the dramatic restrictions on the freedom of movement of skilled workers in England, the ban on these workers emigrating until the 1820s, or think too of German porcelain workers who were simply locked and forced to live in the factories where they worked, so great were the compulsions on the trade secrets involved in the elaborate porcelain processes of the eighteenth century.[10]

The public view of science well into the nineteenth century was not all that remote from the one expressed by Rousseau in his famous *Discourse on the Sciences and the Arts*, in which he lumped together, for example, geometry and poetry, suggesting that both were born of luxury and indolence. 'We have physicists and geometers and chemists and astronomers and poets and musicians and painters. We have no more citizens,' wrote Rousseau.[11] If studying geometry is equivalent, from the point of view of society, to writing Greek verse, then no wonder geometers can travel freely. But once geometry acquires a social and economic importance, then the problems of the geometer inevitably take on a new form. It is worth adding here that it is not simply because of economists and politicians that science is now seen in this way. Members of the scientific community such as Vannevar Bush contributed importantly to the public recognition that science is in some way the source of economic progress. Others, all of us in universities around the world, have profited from the new, post-war, ideology of science.

MILITARY RESEARCH

The problems of the scientific community have been compellingly described elsewhere in this volume. What I would like to do is point to some of the ways in which these problems are related to the economic interpretations I've discussed, and then to look at the implications of the new problems, not only for science policy but also for the economic and other social policies of our societies. But first I want to mention a subject which, to an extraordinary extent, represents a sort of forbidden city in scholarly and policy discussions of questions of science and society. I am talking of military research. Quantitatively and qualitatively, military research looms very large in the worldwide research effort. In Sweden, military research has declined significantly in relative importance in the 1970s. In both the United States and the United Kingdom, military research budgets have increased substantially throughout the 1970s. The United Kingdom alone spends more public money on defence R&D than all the OECD countries together spend on research on environmental protection, transport, and telecommunications.[12] Military research is also increasing fast in France. In the United States military 'research, development, test and evaluation' account for 69 per cent of all research and development to be supported by the federal government in 1984, an increase from less than 50 per cent only four years before.[13]

This funding influences the quality, or the character, of scientific institutions in ways that are at least as important as those that have been discussed in other papers. Think, for example, of the consequences of military funding for the distribution of scientific research by discipline, of the relative emphasis put on the one hand, on physics and engineering sciences and on the other hand, on biology and other life sciences. Think of the effects of secrecy on a large part of the scientific institution. 'Compartmentalization of knowledge, to me, was the very heart of security.' That was the famous observation of the commanding general of the Manhattan Project, and such efforts have continued in some form ever since.[14] Think of things which have a more direct effect on the economic consequences of scientific activity such as the relationship of scientists and technologists to the criterion of cost: if you are working for national defence, cost means something quite different than if you are trying, for example, to perfect a better toaster.

Any realistic account of the pressures on scientists, I suggest, must take account of military institutions. One final note, however, about the expansion in military research over the post-war period. This research, under conditions of great secrecy, does nonetheless seem to suggest that the effort to constrain science within national boundaries is likely to be doomed to failure. Many of the major breakthroughs in military science and military technology over the last generation have, in fact, become widely disseminated among allies and enemies alike. Everything seems to leak, even with the penalties associated with military research, and sometimes in perverse ways. In 1942 Stalin was apparently stimulated to begin the Soviet atomic bomb project by a letter from a young physicist who was an air force lieutenant. The physicist wrote that while on home leave he had gone to the university library to look at physics journals and had in particular looked for articles about spontaneous nuclear fission. He found no articles on this subject, very little of importance about nuclear fission, and he found that all the big names in the field had disappeared from the journals. From this he concluded that the Americans must be working on an atomic bomb and he wrote to Stalin and urged him that the Soviet Union should do the same.[15]

Through omission and commission, other military secrets have continued to leak around the world and this process should, perhaps, lead citizens of those countries which spend a lot on military research to wonder whether their own national security interests are well served by an unending search for technological breakthroughs. I mentioned earlier Robert Fulton, the American inventor who sold submarines to both sides during the Napoleonic Wars. When the British prime minister of the time went to see Fulton's tests of torpedoes and other forms of submarine warfare, the leading British admiral wrote, 'Pitt was the greatest fool that ever existed, to encourage a mode of war which they who commanded the seas did not want, and which, if successful, would deprive them of it.'[16]

THE ECONOMICS OF R&D RECONSIDERED

To return, however, to economic expectations and the new pressures on science. The economic crisis of the last ten years has led to some reconsideration of the basic assumptions – not only about science and economic growth but also about economic growth more generally – which have been the articles of faith of the post-war period. Obviously, those who expected that science, innovation, or something else, would guarantee continuing high levels of productivity growth and employment have been discomfitted. The famous residual, for example, which was thought to explain a large part of economic growth and which in certain theories was identified with 'the advance of knowledge' actually turned negative in the mid-1970s in the United States.[17] Was this therefore to be interpreted as the regress of knowledge?

The point is not that science, which once was good for economic growth and economic progress, has now turned bad, any more than one could say that science, which once led to greater health and progress for mankind, now leads to environmental problems and ill health. Rather the experience of the 1970s and 1980s suggests the limited usefulness of looking at science or at the economy simply in terms of global or macro-economic quantities. Instead, it is essential to consider the differing evolution of individual sectors of national economies and therefore the very diverse ways in which science affects our economy.

This challenge is well illustrated by the evolution of employment over the period of the economic crisis. In all the OECD countries the main growth in employment has come in service producing industries; not in industrial services but by and large either in government or in more or less labour-intensive personal and business services, notably health care. Yet these economic sectors are influenced by scientific change in a way that is quite different from Bush's model of how science leads to new products – reduced costs, increased productivity, and so forth. It is quite different, too, from the model of the industries which themselves perform large amounts of R&D, such as the aircraft industry, chemical and petrochemical industries, even the machinery and transport equipment industries.

In the United States, three industries have created more than one and a half million jobs since 1973 and have together provided over half of all new private jobs created in this period. These three are: eating and drinking places, including fast food restaurants; health services, including private hospitals and nursing homes; and the compendious industry which is called business services and includes personnel supply, data processing, protection and services to buildings, ie, office cleaning and so forth. These three boom industries loom very large in total employment. The increase in employment (two million positions) in eating and drinking places in 1973–81 is greater than total employment in the electronics and steel industries combined. These three together employ more people than the entire heartland of productive industry: mining, construction, all machinery, all electronic equipment, chemicals, and aircraft, as well as automobiles and steel.[18]

These boom industries and comparable industries whether public or private in other countries bear a curious relationship to the virtuous circle linking science and economic growth, and to some extent also to science itself. Their expansion has to a considerable degree been made possible by science. The example of medical services is dramatic. The practice of medicine has been totally transformed by new scientific discoveries and the technologies which embody these discoveries. Yet at the same time the economic consequences of technical innovation in health services are far from classic. A high proportion of the people employed in health services are not working with highly sophisticated equipment. They are looking after people, they are cooking, they are significant way.

cleaning, and so forth. Technology has contributed little to increases in the measured productivity of these services, and in most countries it has not yet contributed much to reducing the costs of health care in a significant way.

The expansion of the business services industry, similarly, was to a significant extent made possible by research, on, for example, data processing and the development of computers, the invention of photocopying machines, and so on. Yet at the same time a large proportion of jobs in this sector in the US and elsewhere are created in occupations such as cleaning and services to buildings. There are more janitors employed in business services in the United States than computer programmers. Once again, the consequences of technical change for increased productivity are far from predictable. The third boom industry, eating and drinking places, is less influenced than almost any other by science in a formal sense. This industry has also experienced a significant absolute fall in its level of productivity in the United States over the period of rapid expansion that has now ended.[19]

What are the implications of these sorts of changes in employment for the relationship between science and society, and for the policies affecting science that these relationships imply? They suggest that there are serious limitations to the basic articles of faith that have governed our view of this relationship over the post-war period. In the case of the first of my four pieces of received wisdom, I think the experience of the last ten years suggests that the causal relationship between the accumulation of scientific knowledge and economic growth is much more complicated than has widely been assumed. In particular, the view that science or education is a quantity assimilable to the stock of land or of capital, is seriously misleading.

In the case of the second proposition, to do with automation, the picture is equally mixed. In some industries, advances in production technologies have indeed led to dramatic reductions in employment. At the same time in other industries labour-intensive techniques have dominated. So some of our societies may be close to approaching a sort of dual economy with the threat of automation very real in one segment and large-scale employment of people without formal skills in another. The third proposition, that science and technology are the proximate cause of environmental problems and a decline in the quality of life is equally oversimplified. In some industries new technologies have indeed posed new threats, not only to the physical but also perhaps to the psychological environment. In others they have provided the solution to such problems. The capacity of science and technology to create new markets for environmentally less demanding goods and services has scarcely been touched.

The fourth propostion, that innovation through its effects on international competitiveness is a major source of economic recovery, also is only partly true. OECD studies show that high technology products, however defined, constitute an important and growing part of the exports of countries such as Sweden whose export performance has been impressive over much – if not all – of the post-war period. At the same time, the present emphasis in many countries on the same high technology industries, largely clustered around biotechnologies and electronics, poses the prospect of intensified competition if and when world export markets recover. And they too pose problems of a potential dual economy with an advanced sector providing very few new jobs and a much larger population producing low-productivity public and private services.

CONCLUSIONS

What are the major practical implications of these complexities? Above all, I think, they suggest that our countries need not more innovation but a different sort of innovation. They may also suggest that we should no longer think in terms of more science *per se* as being an absolute good, or at least an absolute good from the point of view of the economy (I will come to other social points of view later.) for once one is concerned with different kinds of innovation, then the problems of technology policy, of policy for the application of science, become remarkably similar to other problems of national economic and social policy. Questions of the diffusion of innovation, ways of providing information about government innovation programmes to small private and public enterprises, assume tremendous importance. Questions of regional participation, regional access to information, which are so essential to labour market and environmental planning are also critical to planning for technology policies. How does one begin to think about the application of technological innovations towards increasing productivity, improving quality of care and work and reducing costs in the private and public health care services sector, for example? How does one stimulate the kind of qualitative innovation in environmental goods and services that was implicit in the hopes of the early 1970s? How can governments stimulate the participation of workers themselves in the creation of new products and new enterprises?

A study done at MIT for the Swedish Board for Technical Development showed that to an interesting extent it was people with on-the-job training rather than people with high levels of formal education who were responsible for the creation of new enterprises, new exports and new jobs.[20] These people perhaps are more like the British artisans who were forbidden to travel before the 1820s than they are like Humphry Davy or even the inventor of the submarine. How do their energies and their ideas fit into the process of diffusion of innovation? Such problems are messy from the point of view of national policy. They require a knowledge of the peculiarities of individual industries, individual regions. They require participation between public authorities, private industry, unions, public employees, communities, regional authorities. They deny the simplicity of macro-economic policies both in their strictly economic form and as they affect the role of science in economic development. But they are essential to finding a way out of the equally messy crisis of the 1980s.

What conclusions follow from these evocations of complexity and messiness? My conclusion is not that governments should no longer support science. It is rather that they and we should reject the view that science in some relatively simple way leads to economic growth. This view is inadequate. It ignores the sectoral and the human complexities of economic growth. It has failed the test of experience, the experience of the 1970s. It also contains a fallacy of principle directly relevant to the pressures on science and to the view of science as a commodity.

On the Turner–Bush–Kuznets composite view, knowledge or science is comparable to an expanse of wilderness, a stock, a commodity, which, although immaterial, can be accumulated. Yet it is a stock which is at the same time unlimited. This proposition is in itself fallacious. For to the extent that science is, and I quote Kuznets again, 'the product of the free, inquiring mind', it is, by definition, limitless. But to the extent that science is a commodity, is something that can be stocked, that can be assimilated to a factor of production, then it is subject to all the laws that govern economic life; it is not limitless. In this sense, the proposition which states a basic causal relationship between science and economic growth is either a tautology or misleading. To the extent that science is free it is limitless. To the extent that it is a factor of production it is limited, by the economic and social and environmental and other human constraints which determine our daily economic life.

The scientific community has not over the post-war period been energetic in examining and criticizing such contradictions. To do so would perhaps be to examine too closely the goose that lays the golden egg of government support for science and innovation. In the

140

last couple of years scientists have not been conspicuous in questioning whether the recent expectations that high technology will lead us out of the economic crisis are in fact justified. Nor, indeed, have scientists been conspicuous among those criticizing increased allocations of public money for military science. This is striking in the United States where recent budget proposals which would yield an increase of $16.7 billion in military research funding since 1980, compared to a decline of $2.6 billion for non-military federal research funding, have been welcomed as a 'pro-science' budget by large parts of the scientific community.[21] Yet unfulfillable expectations of science as a useful factor of production may harm the scientific community itself. The policies that follow from such assumptions are wrong for science and so they may be wrong for society.

I conclude by returning to the wider social interest in supporting science, to questions that go beyond economics. It is evidently true that society should support scientific research as something that is good in itself. I share the belief that a free scientific community is something that is noble in itself. I suggest, too, that if more of our citizens worked as scientists and fewer as office cleaners the level of well-being and fulfillment in our societies would be higher. So, for such reasons and others, science is something that should be supported in and of itself. But this may not be all that remote from Rousseau's view of the sciences and the arts. And, curiously enough, a Rousseauian view of science could actually turn out to be economically useful. Perhaps we do need a more hedonistic science. Does this mean more biology and chemistry and less physics? Does it mean, as has been suggested elsewhere in this volume, less specialization and more women in science? Perhaps such changes might actually yield a different pattern of innovation and even a different science which will enable us to create a future for industries like health care or restaurants.

The French gastronomer and philosopher, Brillat–Savarin, expected in the 1810s that the great triumph of science in the twentieth century would be to create unimaginably delicious tastes out of stones and minerals. Bacon's own rather austere wise men were actually engaged in creating not only machines but also delicious smells, amazing tastes, seductive imitations of bird sounds, new ways of playing, new sensations of all sorts. Perhaps what we need for our societies, and even indeed for our economies, if they jointly succeed in rising like Phoenixes from the present economic crisis, is something more like a science of, in Rousseau's phrase, luxury and indolence; a science of hedonism, a science for an age in which we can all afford to be indolent.

141

REFERENCES

1. Bush Vannevar 1960 *Science: The Endless Frontier*. National Science Foundation, Washington DC, p 47. A Report to the President by Vannevar Bush, Director of the Office of Scientific Research and Development, 5 July 1945.
2. Turner Frederick Jackson 1920 The West and American Ideals, Commencement Address, 17 June 1914 at the University of Washington. In *The Frontier in American History*. Henry Holt and Company, New York, p 300.
3. Turner Frederick Jackson 1920 Pioneer Ideals and the State University. Commencement Address at the Indiana University, 1910. In *The Frontier in American History, op cit*, p 284.
4. Kuznets Simon 1971 *The Economic Growth of Nations*. Harvard University Press, p 323.
5. Kuznets Simon *op cit*, p 333.
6. Wiener Norbert 1964 *I Am a Mathematician*. MIT Press, p 295–6.
7. Bacon Francis nd New Atlantis. In *The Essays*. Odham's Press, London, pp 317, 337.
8. Salomon Jean-Jacques 1973 *Science and Politics*. MIT Press, p 213.
9. Parsons William Barclay 1922 *Robert Fulton and the Submarine*. Columbia University Press.
10. Landes David 1969 *The Unbound Prometheus*. Cambridge University Press, p 148; Bok Sissela 1983 *Secrets*. Pantheon, New York, p 138.
11. Rousseau Jean-Jacques 1859 Discours sur Les Sciences et Les Arts. In *Petits Chef-d'Oeuvre*. Didot Frères, Paris, p 23.
12. 1982 *Science and Technology Indicators, Basic Statistical Series*. OECD, Paris, vol B, January.
13. 1983 *Special Analyses, Budget of the United States Government Fiscal Year 1984*. Executive Office of the President, Office of Management and Budget, Washington DC, p K–28.
14. Groves General Lesley R. 1959 *Now It Can Be Told*. Harper and Row, p 140.
15. Holloway David 1983 *The Soviet Union and the Arms Race*. Yale University Press, p 18.
16. Parsons William Barclay *op cit*, p 90.
17. Denison Edward 1975 *Accounting for Slower Economic Growth*, Brookings Washington DC p 65
18. Rothschild Emma Reagan and the Real Economy *New York Review of Books*. 5 Feb. 1981.
19. 1981 *Productivity Measures for Selected Industries, 1954–79*. US Department of Labour, Bureau of Labour Statistics (Washington DC, April, p 197.
20. 1983 *The MIT Report* **XI** (4) April.
21. 1983 *Special Analyses, Budget of the United States Government Fiscal Year 1984. op cit*, p K–28.

CHAPTER SEVEN

Does It Only Need Good Men to Do Good Science? (Scientific openness as individual responsibility)

Helga Nowotny

FROM BACON TO BERNAL: DOES IT ONLY NEED GOOD MEN TO DO GOOD SCIENCE?

Ever since its inception modern science has continued to adhere to one guiding utopian vision: that of the ideal scientific community. Faced with the 'greatest publick unhappiness' and 'while the considerations of Men and humane affairs may affect us with a thousand various disquiets', as Thomas Sprat put it in 1667, the scientific community offered 'room to differ, without animosity' and the enquiry into nature would even 'permit us, to raise contrary imaginations upon it, without any danger of Civil War'.[1] It would not only offer new procedures for settling differences by argument and experiment, but the Court of Reason would eventually become the place in which all matters, including the unhappy, public affairs would be adjudicated.

In its essence, it is a community composed of an elite that – paraphrasing Ravetz paraphrasing Bernal[2] – I will call good men doing good science. The good men are those who have espoused scientific rationality as the guiding 'scientific world view', as Otto Neurath was to call it in the 1930s in the wake of a new enthusiasm for planning based upon science.[3] The good men were in possession not only of the weapons against darkness and ignorance which every age seemed to bring forth in a new guise; they were also the carriers of an explicitly formulated claim towards a kind of moral superiority based upon the inherent intellectual superiority of scientific rationality. Contrary to science's official apolitical stance, the better insights that science provided were not limited to the scientific realm properly speaking. Rather, what the utopian programme and its claims offered was their extension into the unruly realm of human affairs. If only scientific rationality were to reign and guide their management, thus runs the deep conviction of the good men doing good science, order would at last be brought into the messiness of social life.[4] The final test would be 'this great enterprise of our time, testing whether men can . . . live without war as the great arbiter of history',[5] ie whether science would in the end be able to supersede the recourse to physical violence as a means of settling differences; and whether, on a less existential scale, a more efficient management of human and natural resources would supersede the wastefulness that now prevails in social and economic relations.

The claim of goodness is therefore a double one: it needs good men to do good science – meaning that in science only the best will achieve – but also that good men doing good science are good at something else: they extend the boundaries of science in the world, the process of seemingly limitless expansion in which science is engaged. In the imagery of the conqueror, the liberating and progressive consequences of this expansion clear the way for a more emancipated humanity in general. This theme of the scientific elite being the best and only guarantee of good science and the good state of human affairs has been relatively unchanged from Bacon to Bernal.

Yet, even when looked upon as a utopian vision, the deficiencies in its claim towards moral superiority are apparent: Bacon's *New Atlantis*, grandiose as it was as a prophetic scheme, remained a fragment. The other part that Bacon intended to write on the Best Imaginable State, was never written. We may also note that Bacon who, as a statesman, knew what catstrophies meant and how they occurred, has carefully eliminated them altogether from his island of scientific utopia. What power and knowledge, after having met in one, would ultimately be used for, remained a question left for his intellectual descendants to answer. Likewise, Bernal has eliminated any traces of incoherence or disorder from his vision of the social function of science. He had the greatness, as Ravetz has rightly reminded us, to perceive clearly how applications of science can be blocked and distorted by commercial greed and to analyse how the cycle that leads to and from human needs towards the application of scientific research can be interrupted, distorted or destroyed by secular institutions. Yet, he was also firm and convinced about the simple solution he put forward: change the social order, which is now unjust, inefficient, downgraded by the vices of rampant capitalism, and you will be able to change science.[6] Today, with the sad benefit of hindsight, we know that this simple change in context of application has not eliminated the abuse of power; rereading the writings of those times we are appalled by the lack of sensitivity to our present predicament of what can go wrong and by the strong scientistic and technocratic elements in what a 'world adapted to man' looks like. For historical experience was to teach us that there was no end to the new problems arising after science had tipped the economies of scale and began to move towards becoming more fully industrialized, bureaucratized and militarized.

To speak about the individual scholar and the scientific community and, within this relationship, about openness as the responsibility of the individual is insufficient. It is simply not enough to look at the human, individual frailties without having a sufficiently clear grasp and understanding of the imperfections of institutions that produce and shape individuals. It will not be possible to change the men who do science without taking into account how science has changed in producing certain types of men (I will revert to the question of women doing science later). Science has extended into so many other areas of life, but, as other conquerers have had to experience, it has assimilated certain traits of the conquered in doing so. The claim to moral superiority and to leadership in human affairs has been severely battered. With inevitable clarity, we know today that things can go wrong: good men can be drawn into doing bad science.

THE ORGANIZATION OF COMPLEXITY: THE CASE OF MOLECULAR BIOLOGY

The organizational model, upon which the premise of good science as the convergence of high standards in moral terms and scientific work alike was based, has undergone a profound transformation. The university, once the traditional home for the systematic production of knowledge and the transmission of learning, has had to make room for its former adjunct, the research laboratory, which has taken on an organizational life of its own. Recent studies in the sociology of science have prided themselves on discovering 'life in the lab' as the 'factory in which order is produced'.[7] As in other factories in which capital investment is taking place, the operation is geared towards production – in this case, of scientific facts which have to be stabilized, ie standardized, and to be invested with credibility in order to be traded, sold, appropriated. The scientists working in this factory are said to be obsessed to a certain degree with the economic categories of success: they speak in terms of the costs they incur, they think in terms of cost-benefit analyses, they bargain with each other in the 'manufacture of knowledge' and they behave like bankers who control budgets and balance accounts, payoffs and tradeoffs alike. The rules that dominate the research game are in their turn dominated by the economic laws of investment and return, of profitability and success.[8]

Yet, comparing these studies with those that have come out of industrial sociology, we may wonder why it has taken so long to discover a certain convergence in the games that scientists play in the laboratory and the rules that govern behaviour on the industrial shopfloor. Behind such an analogy lies a process of convergence of new principles of organizing complexity in science and in society alike. As an illustrative example of how such convergence began to take hold, consider the role played by the Rockefeller Foundation in the 1930s in the establishment of a new field that would eventually become molecular biology.[9] A new network of informal personal contacts in an as yet unestablished field was created among its upcoming young elite who were to be sufficiently open to the transfer of new methods of management taken over from industry. 'Managed science' meant a goal-oriented approach in supporting a specific theoretical programme, that is, the reductionist programme within biochemistry and genetics; generous financial support for projects which included the improvement of technology as an integral part of and precondition for theoretical advances; support and encouragement for a collaborative research style and the expansion of a subtle, yet highly efficient, system of patronage designed to integrate young and promising talents into the new support programme. The deeper significance of the process of convergence between theoretical and strategic-institutional programmes can be grasped by analysing the meta-language which was developed around it: in it certain key concepts were used both for the analysis of nature and its societal appropriation; with formal methods being used in analysing living organisms as well as in organizing research.

The success of molecular biology by and large proves the efficiency of the organizational model that contributed to it.[10] The 'capitalization of life', as Yoxen has called it, became the new basis for biotechnology. Life itself is now seen as programmed and programmable and thereby undergoes a redefinition: it can be appropriated as intellectual property, it can be traded, bought and sold. Science as a commodity has taken on a new form and information becomes one of the keys for the organization of complexity.

THE DISCOVERY OF TECHNOLOGICAL SWEETNESS: THE LEGACY OF THE MANHATTAN PROJECT

The rationalization of the research process in accordance with new management principles and the converging organization of complexity, both on the theoretical-programmatic legal and in the comprehensive sense of organizing training and mobility requirements in a multinational chain, spanning industry and universities alike, provide the wider setting of the forces that have decisively shaped the exterior conditions of what has been called the industrialization of science, or more recently, its 'collectivization'.[11] The internal changes, however, were no less marked: as the most far-reaching transformation they signal the change from the scientist working essentially as an individual to scientists working in a group. The new form of competitive cooperation which resulted from this seemingly innocent shift had a profound impact on the individual scientists' motivation, reward and career structures, self-image and sense of responsibility. The unifying process which directed this transformation was the shift from the production of discipline-based knowledge to the realization of science-based, technological projects – from science to science-based technological research.

Thus, what has somtimes been described as mission-oriented research is a much more profound change in the organizational mode of doing research than some have realized. Its history, as some of the metaphors still used convey, is war; its ancestor is the Manhattan Project.[12] There is not the space here to go into the detailed history of that project. Rather, I want to concentrate upon some of its structural legacies that appear to have come to stay. In the aftermath of its 'successful termination' – the dropping of the A-bombs – a more or less sincerely felt and profound moral discussion set in. What was now problematic were the altered conditions of the application of scientific knowledge; the loss of innocence of science; the changing relationship between the scientific and the political elites; but above all the moral implications for the individual scientist.

What was not discussed, since it was not perceived as being problematic at the time,[13] were the side effects of the lasting organizational changes that turned out to be satisfactory and beneficient to both sides: to the scientific community and to the political–military establishment alike. The new mechanisms of scientific collaboration and masterminding science-based, technological research on an unprecedentedly by large and complex scale turned out to be too efficient not to remain. This efficiency has been refined and extended even though the particular historical conditions that gave rise to it have ceased to exist. Indeed, these mechanisms have been adapted to a smaller scale and they account to a large degree for the 'new breed' of scientists that we complain about as having lost openness.

Let us take a second look at what is known. One of the ingredients of the success of the Manhattan Project consisted in bringing together a group of highly qualified scientists and engineers from different backgrounds and disciplines – theoreticians, experimentalists, industrial engineers – and integrating them into a functioning, collaborative team. The towering influence of a strong personality was a decisive factor. Other ingredients of the successful model were: the strong moral and political appeal, which meant working under an urgent moral verdict to beat Nazi Germany in the race for superior weapons technology; the tremendous scientific interest aroused by the underlying theoretical problems; the fact that a segment of the scientific world elite was able to work together under unusual, immensely crowded but apparently satisfying conditions; the unquestioned disposal of almost limitless resources in highly concentrated form; and, perhaps most important, the direct concentration upon a specific task with a practical outcome: the bomb.[14]

Once this newly invented organizational model succeeded, admittedly under somewhat unusual conditions, it was too beautiful to let go. A number of institutions that were founded in the United States after the war incorporated the new organizational model, although there were heated discussions about the concrete form it should take: it was conceded that a certain amount of steering from the outside was both necessary and beneficial, but the well-kept illusion of autonomous scientists inside was also strongly maintained and the public funding of research became the determining factor of the new science policy. In its more mundane form, task-orientation shed some of its military armour and blended well with the glamour of well-organized science and technology.

The superiority of the new organizational model lay precisely in its capacity to integrate different types of researchers, different skills and experience into some kind of optimal mix – always defined in view of a relatively short-term, concrete and practical task. The necessary correlatation of increased individual mobility implied new career structures, in the sense of being able to move on to the 'better teams', ie the team with even more qualified researchers, more interesting problems and better-funded tasks. The inherent rewards were no longer defined simply by a reputation in one's own discipline, achieved, mainly, through discipline-based publications, but shifted towards the satisfaction of achieving a practical outcome, of being associated with a successful project. It became the company that lent its name to the product and being part of the company became an intrinsic reward.

But there were further repercussions of the Manhattan Project. Standing on the shoulders of giants and making one's own small contribution to the roster of claims towards immortality – which had been the aspiration of individual members of the scientific community for a long time – was also the hallmark of continuing a certain intellectual tradition. It was the hallmark of what I shall call 'noble science'. By contrast, the new scientific entrepreneurs, the condottiere of project-oriented science, were attracted by what one of their greatest representatives aptly called the discovery of 'technological sweetness': the science-based, interdisciplinary orientation, set up and geared towards realizing a technological project that works. Its 'sweetness' conveys an attachment to the product as such. It is not seen in any larger context of application, but for its own, practical sense and which displays the quasi-erotic features of the automaton brought to functioning or life by the (male) scientist himself. It conveys something of the sensual pleasure that comes with the sense of total control, set up at the expense of narrowing the context in which the product is envisaged and made to work. Finally, sweetness contains also the invitation to try out, the offer of consumption – with a faint reminiscence of the paradisical apple.

Even if we know that the overwhelming majority of practising scientists and engineers worked and continue to work in conditions far removed from condottiere glamour, the appeal of technological sweetness is pervasive: we have but to take a glimpse at what Turkle in her studies of computer scientists has called the subjective side of the computer[15] in order to understand the intricate, compensatory mechanisms that constitute the technological appeal of their work routine which becomes transformed into something adventurous, non-routine and extraordinary when invested with the fantasies of omniscience.

SCIENTIFIC OPENNESS – A DISAPPEARING VIRTUE

We need not reiterate here the fundamental contribution that scientific openness has made to the development of modern science; its inherent value in both a scientific and in a political-democratic sense. The example given by Ravetz – that it took a student to discover that the niobium used as an alloy in the steel of pressure vessels in a pressurized water reactor has such intense and long-lasting radiation that the decommissioning of such reactors will be enormously more difficult and expensive than previously assumed – cannot only be taken to ask what the scientific experts were doing who were supposed to check on such a possibility during the previous decades, but it is also a testimony to the still functioning openness in our institutional set-up, even in the realm of technology. Yet it may become a rare example, if the appropriation of intellectual property, its fragmentation and commercialization continue to be drawn into military secrecy by stages for the military domain already occupies almost 50 per cent of the world's total scientific and technological labour force.

It is worth taking yet another look backwards and recalling what some of the structural changes were that brought about the emergence and valuation of scientific openness. Initially, the norm of openness constituted a radical break with the secret tradition of the magus's trade. In Bacon's *New Atlantis*, the Mercatores Luci, who went out every twelve years in order to gather new scientific information of potential interest to the House of Salomon, went *incognito*. In another passage, we are told about the deliberations undertaken in order to decide which of the discoveries and knowledge of nature are to be published and which ones not. The members were bound by an oath of secrecy and although some things are revealed to the state, others are not.[16] We can assume that the economic context of incipient trade capitalism encouraged the free exchange of goods and the free exchange of scientific information alike, but the practice probably took longer to become established. It meant that the narrow and conventional boundaries of the craftmen's guilds and their inward-bound communication structure had to be transcended, and this occurred only with the rise of the nation states. As long as the norms and behaviour of free exchange, in the widest economic sense, were thought to benefit the national interest, openness was encouraged and the universalist aspects of the message of the Enlightenment would receive enthusiastic support: the ideological and utilitarian components of openness would meet.

Today, with the new information technology at our disposal, scientific openness is challenged even on technological and social grounds. It is not merely that since science has become a commodity, so scientific information also has turned into a commodity to be bought and sold, guarded and destroyed. Access to it can be regulated and controlled in unprecedented ways and the impact of the new information technologies in doing so has been unduly neglected. It is not only secret information within the military research sector which has been separated and withdrawn from the ordinary network of scientific knowledge diffusion and circulation. The establishment of large-scale information systems with selective access implies a potentially deep structural change that may lead to a two-class system in access to scientific information.[17]

Faced with these and other developments which threaten scientific openness, what can we say about the responsibility of the individual scholar confronted with the decline of a disappearing virtue? First, we have to realize that virtues are not randomly distributed, nor are they something inherent or naturally given. Rather, they are tied to the emergence and decline of different social groups. Up to the Second World War science resembled in several of its structural features the former nobility: in its most visible activities at least, it was the preoccupation of a tightly inter-locking small elite, spread out across national boundaries and highly respected in its devotion to the pursuit of systematic knowledge and truth. Openness was a virtue linked to science as a noble activity. With the gradual and relative decline of science as such an activity, its corresponding virtues are also threatened and may disappear. For the new rising group engaged in it, the condottieri of project-oriented research, openness is dysfunctional. Competition between such groups gives advantages to those who know how to exclude others from information more often than by sharing it, at least at certain critical stages. Since publications – the former regulatory mechanism in establishing claims of priority and reputation and a strong structural support for openness – are also becoming obsolete or at least secondary in project-oriented research, legal regulations will eventually replace the self-regulation that has prevailed so far within the scientific community. In a curious reversal of Bacon's prophecy, the state will decide what is to be published and what not. In addition, with huge data banks becoming progressively more available, scientific information becomes – from the point of view of the individual scholar – externalized and reified in the sense that no individual can any longer keep up, let alone master the flow of information necessary for the pursuit of research. It is taken over, like many other activities in daily life, by a huge impersonal apparatus governed by rules of access and retrieval, thus making it easier for the individual scholar to accept a contraction of openness.

But with science and technology occupying the place in the world they hold today, scientific openness has acquired a new and additional meaning. I refer to the changing relationship between science, technology and the public. With the actual decline of scientific openness as an internal regulatory mechanism, the ascendency of openness as an external relationship directed towards the public may even gain in additional importance and urgency. In the last decades we have witnessed controversies around large-scale technological developments and we are about to undergo another deep crisis about nuclear weapons which undoubtedly will affect the public's attitude towards science and technology in general. Openness in this field of conflict does not mean to engage in the kind of pseudo-public relations activities that have served as an inadequate surrogate in the last few years. Rather, it is a serious challenge, calling for an institutionalized pattern of honest and open exchange of information on matters of legitimate public concern. I think it is in this area where there is still much room for individual responsibility and collective responsibility alike. Futhermore, openness in this second sense is not only compatible with the new style of scientific research, but becomes mandatory if this style is to survive a growing distrust of the social and political consequences of the results of this kind of research. It is therefore also in the self-interest of science as project-oriented research to expand its responsibility in the direction of working towards a more trustworthy public basis.

Scientific openness, even if it may be a disappearing virtue that will fade eventually together with the world of noble science, may remain with us in a broader, more democratic and urgent sense.

CAN GOOD MEN FAIL?

We have come full circle to our present predicament which is that we know that in science and technology too, things can go wrong. The old tenet of noble science was that individuals can fail, but that their errors, their frailties, even their personal weaknesses, will be compensated by the collective march of scientific progress. The self-correcting mechanisms within science as an institution were deemed to be sufficiently strong and efficient to guarantee that despite minor aberration the deviation on the route to progress would never become sufficiently large to allow a serious derailment. However, this faith in a collective venture which was rooted in the working experience of academic scientists, can no longer be maintained in face of fragmentation into highly mobile, decentralized and transient research units, which are constituted and dissolved in accordance with the life cycle of a multitude of projects. We have come to realize that it is neither sufficient to presume that good men alone will do good science – in all the various meanings of the word – nor is science alone capable of producing good men whether inside or outside science. And, anyway, where are the women?

Would science be any different, if women were included in sufficient numbers, would it have a more human face? Like any hypothetical case it is easy to argue and impossible to prove. I must admit, however, that it is hard to imagine that women would be as easily seduced as men by technological sweetness; that they could as readily be drawn into the 'alchemy of the arms race' as S. Zuckerman has called it; that their blindness *vis-à-vis* potential global disaster would be identical to the blindness of their male colleagues, rather than be of a different kind. But it can hardly be expected that the presence of women in positions now occupied by men would of itself change the structure of science. The rules of the present game would have to be changed radically. But even then, good women – even if they have other, not yet tapped human resources than men – are also human; they too can fail.

Where does this leave us finally? If we ask what has happened, the answers are easy to give: technological sweetness has turned sour; war, although a sad constant in the conduct of human affairs, has become an apocalyptic vision with the help of science and technology and what Ravetz calls technological blunders are becoming increasingly menacing. It is much more difficult to answer the question why. There are no obvious scapegoats to blame, no single agency or institution, let alone groups of individuals. Rather, we are confronted with a large-scale historical process, a blind historical process, directed, but not planned as such, in which we all participate. In its unfolding, science has gained the means of changing, even annihilating the world. As other monopoly holders have experienced in other fields, the successful acquisition of the monopoly has changed the nature of science.

It is the transition from science as a noble, academic activity, which was mainly university-based, to a science-based technology and a project-oriented research-science that has been the focus of my exposition. In this process of transformation, scientific openness in the traditional sense is likely to fade out, while being replaced by scientific openness as a new and yet to be institutionalized challenge. Good science and good men are insufficient. Good women have had no chance, as yet, to alter the rules of the game. Nor is it sufficient to believe any more that only the social and political order has to be changed if good science is to have a chance. The convergence of scientific activity with profit-orientation, the bureaucratization and militarization of science have been the topic of this book. But an appeal – even a strong moral appeal – to individual responsibility alone will not do either.

If good men can fail – which is a very likely possibility – what we need are better organizations. This means deliberate designs, constructed with a view to the traits we value and wish to preserve. It is a call for our collective imagination to envisage what such organizational designs could look like, based on a thorough understanding of the functioning off organized scientific life operating within a social and political context. Nor is this a task unique to science. After all, it has taken centuries and many struggles until political institutions have been devised that allow for the play of checks and balances that constitute the central tenets of our democratic institutions.

One important question still remains, however: how can organizations know what is good science? Or would we opt for less?

ACKNOWLEDGEMENTS

My sincere thanks go to Wolfgang Reiter who spent a long, sunny Sunday afternoon with me in a Viennese coffee-house, discussing the Manhattan Project and its aftermath.

REFERENCES

1. Sprat Thomas 1667 *The History of the Royal Society of London*.
2. Ravetz J. 1982 The Social Functions of Science: a commemoration of J.D. Bernal's vision *Science and Public Policy* October, pp 262–266.
3. Neurath Otto 1979 *Wissenschaftliche Weltauffassung, Sozialismus und Logischer Empirismus* Rainer Hegselmann (ed). Suhrkamp, Frankfurt.
4. Mendelsohn Everett, Nowotny Helga (eds) 1984 Science and Utopia *Yearbook in the Sociology of the Sciences*. Reidel, Dortrecht, vol 8.
5. Oppenheimer Robert, quoted in Nuel Pharr Davis, *Lawrence and Oppenheimer*, New York, 1968, p 16
6. Ravetz J. 1982 *op cit*.
7. Latour Bruno, Woolgar Steve 1979 *Laboratory Life, The Social Construction of Scientific Facts*. Sage Publications, Beverly Hills.
8. See my critique of these studies and further ideas on which this section is based. Nowotny Helga 1982 Leben im Labor und Draußen: Wissenschaft ohne Wissen? *Soziale Welt* **33** (2) pp 208–220.

9. See especially Kohler Robert 1976 The Management of Science: The Experience of Warren Weaver and the Rockefeller Foundation Programme in Molecular Biology *Minerva* **14**: pp 279–306 and Yoxen Edward 1981 Life as a Productive Force: Capitalising the Science and Technology of Molecular Biology. In Levidow Les, Young Bob *Science, Technology and the Labour Process*. CSE Books, pp 66–122.

10. A recent interesting reassessment of the role played by Warren Weaver's management philosophy corrects details, but does not, in my opinion, contradict the thesis advanced here. See Abir-am Pnina 1982 The Discourse of Physical Power and Biological Knowledge in the 1930s: A Reappraisal of the Rockefeller Foundation's Policy in Molecular Biology *Social Studies of Science*, **12** (3) August: pp 341–382.

11. Ziman John 1983 The Collectivisation of Science. Bernal Lecture to the Royal Society, April (mimeo); which came to my notice only after this paper was written.

12. There exists, of course, a vast literature on this subject. Among others, see Kevles Daniel 1971 *The Physicists*. Vintage Books, New York; Grabner Ingo, Reiter Wolfgang 1982 Ende und Fortschritt der Physik. Vienna, (mimeo).

13. This is not to say that research policies were not discussed; quite the contrary. However, the main aim was how to advance the successful organizational model and not to reflect upon its internal consequences. See Kevles Daniel *op cit*.

14. Some have referred to this as a highly pressurized monastic life. According to Daniel Kevles, 'Los Alamos scientists skied and hiked in the remote beauty of the surrounding mountains, gathered at intellectually stimulating social evenings and kept the maternity wards busy'. *Op cit*, p 330.

15. Turkle Sherry 1982 The Subjective Computer *Social Studies of Science*, **12** (2) May: pp 173–206.

16. Bacon Francis nd New Atlantis. In *The Essays* Odham Press, London, p 411.

17. A case in point, is the Advanced Research Project Agency, access to which is open to certain institutions in NATO countries. On the whole, the 'information war' going on behind the closed screens of computer terminals has been neglected by social studies of science.

POSTSCRIPT

Useful Science and Scientific Openness: Baconian Vision or Faustian Bargain?

Björn Wittrock

SCIENCE AS A COMMODITY: THE PROBLEM RE-EXAMINED

Is science just a commodity? Is it something which is produced in response to demands, something which can be bought and sold, traded and stolen? Is its real value determined by the nature of these types of exchange relationships? And if science is mainly a commodity among other commodities, what about science as the epitome of free and unbounded inquiry? What about the notion of scientists as constituting a community of scholars who take part in the search for truth, who freely share their findings and openly expose their results to the scrutiny and criticism of others irrespective of their national or commercial loyalties? The title of this volume – *Science as a Commodity: Threats to the Open Community of Scholars* – might evoke the image of a scholarly community beseiged by societal forces hostile to the very essence of scientific inquiry. Science reduced to a commodity signals the triumphant onslaught of these forces. But if nothing else, this volume should have thoroughly dispelled such a simplistic view. Science has not become a commodity; it always was. Commoditization does not threaten the scientific community; it has consistently served as the basis for science as a professional – and thus autonomous – enterprise. If anything, or so some contributors would have it, it is science unbound and translated into technology – often enough destructive – which threatens human civilization and society, not the other way around.

Science might have become increasingly involved in matters of national defence, public policy and industrial innovation, and this involvement inevitably has repercussions for science as a scholarly endeavour and in particular for academic science as the epitome of open scholarly inquiry. However, this type of involvement has been

156

part and parcel of the emergence of professional scientific communities since the late 19th century. External threats do not simply impose themselves upon pristine communities of academic scholars. Rather, important segments of these communities have long been accustomed to trading knowledge and professional expertise for resources, recognition and professional autonomy.

The real issue is not whether such interactions occur, whether such bargains are struck, whether science is sometimes looked upon as a commodity and as a tool. In fact – as shown by several contributors to this volume and in most detail by Michael Gibbons – when Francis Bacon expressed the programmatic vision of what was to become the Western scientific ideal in *New Atlantis*, the scholars he describes do not pursue an Aristotelian quest for 'true certain knowledge of causal necessity'. The search for 'the knowledge of Causes' is directed at 'the enlarging of the bounds of Human Empire, to the effecting of all things possible'. It is an active and experimental science aiming at producing practically useful results. Furthermore, the Baconian community of scholars in 'Salomon's House' might be a tightly knit and well organized collegiate body, but the flow of information from that body is far from free. It involves careful screening and 'consultations, which of the inventions and experiences which we have discovered shall be published, and which not: and all take an oath of secrecy, for the concealing of those which we think fit to keep secret: though some of those we do reveal sometimes to the state, and some not'. Not only is the dissemination of information severely restricted, but the image of science in *New Atlantis* is certainly consistent with a commodity-oriented pattern of information collection. Twelve fellows of 'Salomon's House' 'sail into foreign countries, under the names of other nations, (for our own we conceal); who bring us the books, and abstracts, and patterns of experiments of all other parts. These we call Merchants of Light'.

The real issue is rather whether the whole configuration of interactions between scholarly communities and a multitude of governmental and industrial groupings is now undergoing change on such a scale that qualitatively new features are emerging in universities and other bases of research and scholarship. Even if interactions are more accurately characterized as bargains than as threats, they might still signal such wide-ranging changes that the very notion of scientists as communities of scholars has to be reappraised. Are we now witnessing such deep-cutting intrusion by governmental and industrial interests into the private lives of institutions of research and learning that it is all but obvious that the notion of an open community of scholars is hopelessly anachronistic both as a normative model and as a reasonable approximation of the actual operating mode of contemporary academic research systems?

SCIENCE INTO POLICY, POWER AND WEALTH: NEW UNCERTAINTIES

Universities and other institutions of higher education and research have always had to respond to challenges and demands in order to successfully uphold their claims to economic support and academic freedom. A steady supply of knowledge and qualified manpower has been the best way of safeguarding the terms of trade for universities and similar institutions. The essays in this volume indicate that this is still very much the case. However, several of the contributions also go one step further; they seem to indicate that research institutions tend to become enmeshed in patterns of interaction of increasing complexity and ambiguity in terms of their long-term implications. Emerging complexities can be traced on several levels.

The promise of rich economic returns from investments in research was a chief impetus behind science policies in the Western world throughout the 1950s and well into the 1960s. The economic turbulence of the 1970s and 1980s has been conducive to a revival of such sentiments. After a decade of sluggish economic performance, high unemployment rates and intractable budgetary deficits many Western governments expect science and technology to show the way out of low growth and stagnation. However, current difficulties do not only serve as an impetus for the support of science and technology. They also act as an impediment to such support. Firstly, financial strictures make it tempting for governments to cut down on inherently uncertain budgetary items such as research. Secondly, the lure of renewed economic expansion entails a temptation for governments to deflect growing shares of research resources into programmes which promise relatively rapid technological returns. Furthermore, there are striking similarities in the list of science and technology priorities of different countries. But – as highlighted by several of the essays in this volume and most distinctively, perhaps, in those by Emma Rothschild and Jean-Jacques Salomon – is it really realistic to expect that there will be only winners in a scramble for preeminence and excellence in fields such as biotechnology, microelectronics and communication systems?

The wide-spread and strenuous efforts to link up science and technology policy and to speed up the translation of technological innovation into economic recovery and growth help channel resources into the scientific enterprise. The contribution by Hermann Grimmeiss provides a lucid explication of the rationale for the support of high-technology industry to free-ranging university research. Keen competition forces high-technology industry to focus on application-oriented research and makes it increasingly dependent on high-quality university research of a basic nature to lay a foundation for the handling of long-term problems. Thus there is a case for more intimate university-industry cooperation which should be mutually beneficial and fully compatible with traditional academic notions of freedom of research. The cooperative arrangements which emerge out of university-industry interaction might more properly be described in terms of bargains than threats. However, there is obviously no guarantee that they will not involve threats to traditional academic perogatives, eg in the realms of appointments and publication of results.

Several contributors touch upon this possibility, and Dorothy Zinberg closely reviews American examples of university-industry deals in recent years. She clearly brings out the constructive responses on the part of universities to potentially harmful intrusion, but she also indicates that there might be an element of differential and preferential treatment involved; the most prestigious research universities might well stand a better chance to uphold traditional scholarly virtues and norms than their less well-endowed brethren further down in the academic pecking order.

Sheldon Rothblatt's historical analysis provides analogous examples from 19th century Britain. This case, however, also highlights the growing resistance to the market in late Victorian high education institutions. This resistance was spearheaded by Oxford and Cambridge and became an established tenet of the major political parties in Britain. The recent responses by British universities to a series of questions from the University Grants Committee indicate the persistance throughout the 20th century of anti-market sentiments in higher education institutions. One of the statements by Oxford University is a good case in point:

'Universities jealously guard their academic freedom and would deeply resent any attempt by Government to restrict that freedom through financial pressure. On the other hand, universities are almost completely dependent on public funds, whether through the UGC, the research councils, and government sponsored research, or student fees reimbursed

by local education authorities. Significant financial independence from
public funding (particularly in regard to general university expenditure)
would mean greater dependence on the market place and much uncertainty
and disruption in academic activities. Such a switch would require a major
change in social attitudes in this country towards the funding of higher
education. It is national policy, rightly in our view, to support higher
education from public funds.'[1]

True enough, interactions between industry and universities have
intensified in recent years in most Western countries. Still, however,
in terms of resources for university research such interactions are
relatively modest in most of these countries and the developments
underway can in many ways be seen as a form of recovery after some
decades of declining industrial involvement in the research undertakings of higher education institutions. Nor have these kinds of
support so far presented major political dilemmas. Thus Swedish
social democrats and British conservatives appear to be equally eager
to promote the transformation of their respective societies from industrial to high technological information societies.

Some participants in the Stockholm symposium, most emphatically
perhaps Richard Stankiewicz, argued that university-industry interaction is essentially a function of the role of science in processes of
technological innovation. A recurring pattern can be discerned in
which random interaction is replaced by 'hot' phases to be succeeded
by 'cool' phases of interaction. In the 'hot' phases basic scientists
become directly involved in processes of technological innovation.
However, these phases are relatively brief. In the ensuing 'cool' phases
new fields of applied science and engineering develop and potential
conflicts arising out of direct interaction between fundamental science
and industrial innovation can be easily kept at bay. From this
perspective what is now happening in the biological sciences is only
another example of a transitory 'hot' phase. The current focus on the
potentially pernicious effects of intense university-industry interaction
is thus misplaced and previous cases such as chemical and nuclear
technology should help dispel fears that the 'hot' phase will prove
permanent. Even if there should be an increased frequency of 'hot'
phases there is, so the argument goes, little or nothing to suggest that
this phenomenon in and of itself should present any insurmountable
challenges to basic research or academic freedom, given reasonably
competent university administrators and sufficient scholarly
self-esteem. The real issue might – in the words of Sheldon Rothblatt
– rather 'be the capacity of scientists, scientific organizations and
universities to maintain a sense of their overarching historical and

professional goals in the teeth of pressures and temptations which must, in the very nature of human affairs, be always present if not in the same form or degree'.

However, it is this very capacity that is seriously questioned by several contributors to this volume. Thus even if the 'hot' phases of university-industry interaction are transitory, some of their most important effects are lasting. Over the course of time they give rise to a growing number of professionally and technologically oriented fields within universities and higher education institutions. Such fields tend to be less dominated by the universalistic values of academic science and more by considerations of their usefulness for particular constituencies and for the solution of problems of design and innovation. Furthermore, the expansion of higher education all over the Western world during the 1960s was often, if not always, accompanied by efforts to emphasize the role of professional and vocational training as well as the responsiveness of universities to government demands for usable knowledge.

Jointly and over a considerable span of time, these types of external and internal processes have come to effect a profound change in the intellectual landscape of higher education institutions. There has been – to quote Helga Nowotny – 'a certain convergence in the games that scientists play in the lab and the rules that govern behaviour on the industrial shop-floor'. The successful mobilization of science in mission-oriented research groups which realized technological, and often enough military, projects – 'technological sweetness' in Helga Nowotny's terms – has gradually changed the ethos of scientific work. The noble men of traditional academic and international science have been replaced by acquisitive and rapacious 'condottiere' of project-oriented science. Thus technological sweetness has turned sour, and higher education institutions find themselves – in the words of Peter Scott – groping for 'a meta-language that is more than technical and administrative and which can impose a moral structure on their exploding experiences'.[2] In the absence of such a structure it is, Peter Scott argues in another passage, 'increasingly difficult to regard the modern university as in any sense an organic academic society rather than simply as a shared bureaucratic environment'.[3]

Bacon's vision in *New Atlantis* was based upon the premise that the scholarly pursuit of the noble men of 'Salomon's House' would lead to immediately useful results. The universalistic norms of science and the existence of a free community of scholars were seen to be fully consistent with the achievement of practical objectives of the emerging nation states. But, as underscored by several contributors to this volume, the Baconian vision did not really materialize until the 19th century. At that time science-based technologies came to deeply affect industrial developments and higher education institutions came to be infused with new types of norms. No longer could specialization and scientific originality be disregarded as aberrations and as detrimental to the collegiate communities and the primary obligations of universities to provide a broad and gentlemanly general education.[4]

During the 20th century and particularly after the Second World War, governments have increasingly tried to link up science to national aims and public policies as well as to tap it as a source of economic growth. The role of science as a commodity and tool has brought the scholarly world infinitely greater resources in the period after the Second World War than ever before in its history. But it has also tended to undermine its identity and its guiding norms. The translation of science into power, policy and wealth can more adequately be described in terms of processes of bargains than of threats. But the bargains have involved a Faustian element. They have not only provided sufficient leverage to permit large-scale research projects. The entrepreneurs and 'condottiere' which epitomize these undertakings do not conform to the noble norms of an international scientific aristocracy, and the cherished value of openness has been intimately linked to the existence of such an aristocracy. Instead, competition between short-lived project groups is conducive to disruptions in the free flow of information and to efforts to exclude others from it.

Thus the supposedly new threats to the open community of scientists in the academic research system may be neither new, nor threats in any conventional sense. Sheldon Rothblatt observes that 'the marketability of science' and 'the desire of powerful industrial states to employ science in the service of national aims . . . may virtually be accepted as constants, and scientists have always been two minds about them'. However, in the last two decades efforts have been intensified all over the Western world to use research as a limitless factor of production to promote growth and to pull economies out of slump and stagnation, as a cure-all guide to policies for social betterment and welfare, and as a rational basis for policies to trim bureaucracy and increase efficiency. Not only may these efforts have highlighted an inherent tension in an established pattern of accommodating universalistic norms and particularistic commitments in modern universities and research institutions. These efforts have also revealed tensions and inconsistencies in many of the fundamental conceptions underlying science policies of recent years. In this volume

Emma Rothschild clearly demonstrates the serious limitations which apply to the basic belief of most Western governments in the post Second World War period that science is the main source of economic growth. Scientists and many governments alike have tended to advance a view according to which science is both free, and hence limitless, and a major factor of production, and hence limited. The point is that 'unfulfillable expectations of science as a useful factor of production may harm the scientific institution itself'.

In a similar vein the 1970s and 1980s have witnessed strenuous efforts in most of the OECD member countries to bring the social sciences directly to bear on the formulation and implementation of public policies and administration in all spheres of society. These efforts have been helpful in generating research contracts to social scientists. Unfortunately, however, underlying many of these policies has been a simplistic belief that well-defined pieces of knowledge can be straightforwardly requested and subsequently fed into a streamlined planning machinery. A large number of studies of processes of knowledge utilization have made it abundantly clear that this conception presents at best an oversimplified and distorted picture.[5] Not only do social sciences risk suffering in the wake of inevitable disappointments over the limitations of their contribution to policy amelioration; there is also the accompanying risk that their real, albeit often but not always more indirect and long-range, significance for policy-making is disregarded in heated controversies between over-optimistic technocrats, disillusioned politicians and academic traditionalists.

Jean-Jacques Salomon notes that Bacon's vision in *New Atlantis* is 'the dream of a science-based society in which knowledge is useful and efficient, in which researchers fulfil functions directly oriented towards utilitarian purposes'. In this visionary society we find 'fundamental researchers as well as engineers, scientific attachés as well as managers, businessmen, diplomats, soldiers and spies'. These developments might well signify 'our farewell to science conceived as essentially a learned and innocent activity, free from the pressures of the powers-that-be'. However, the Baconian scholars of 'Salomon's House' were both learned and useful, aristocratic and merchant-like, committed to unravelling the truth of nature but also changing the face of the earth. The unfolding of the Baconian vision in the intellectual and societal landscape of the mid and late 20th century has involved notions of science as an 'endless frontier', as the true generator of wealth, and as an all-powerful tool of the modern welfare-warfare state. The contributions to this volume have highlighted the inherent tensions and the growing counterfinalities of this vision.

The Baconian expression of the Western scientific ideal leaves room for both openness and secrecy, for entrepreneurs and noble men, for science as truth and as a commodity and tool. In its fundamental societal harmony, *New Atlantis* is a true expression of the utopian tradition. In a world where science plays perhaps for the first time the type of interventionist role envisaged by Bacon, where his vision of a science which changes the physical nature, brings forth new and carefully manipulated biological creatures, and guides statesmen in public affairs, yet preserves a measure of aristocratic detachment, is far less of a vision than a reality, deep-seated ambiguities and tensions inherent in this vision emerge more clearly. But even if the contributions to this volume are remarkably consistent in their analysis of the origins and features of challenges to contemporary science policies, they are refreshingly unpretentious and cautious in proposing easy ways out of the dilemmas discerned. In fact, the outlines of various solutions are mainly indicative and match the complexity and uncertainties of the problems involved.

CONFRONTING THE CHALLENGE TO SCIENCE POLICY

None of the contributions to this volume espouses a *traditionalist defence of the purity of academia*. Science has become a central tool in politics and economics and, in the words of Jean-Jacques Salomon, the 'image of the learned scholar, the "savant" involved in research for its own sake or for his own pleasure is an image of our culture, not a reality in our societies, concerned as they are with the exploitation of research results. Science has become a technique among others and scientists fulfil functions that are today as far from those of the Greek philosophers as an abacus is from a microcomputer.' Thus any realistic assessment of contempory science policy must recognize that academic research has grown and thrived as a result of a multiplicity of exchange relations. Just preserving academic purity is simply not a feasible policy option. But if so, much the same should apply with equal or greater validity to the *nostalgic dream of retracing the origins* of a 'golden age' when scholars constituted truly open communities of 'noble men'.

Although such a notion is found, for example in the essays of Michael Gibbons, Jean-Jacques Salomon, Emma Rothschild and

Helga Nowotny, it is, not surprisingly, in most instances to demonstrate how hopelessly naïve and dated such ideas are. However, as pointed out by Michael Gibbons in his introduction, in a number of the contributions to this volume there is 'an undertow, not always fully articulated, that with the advance of the scientific revolution something was definitely gained but also something was lost and that in general both science and society are somewhat worse off for it'. Maybe the demise of an Aristotelian conception of science meant that something fundamentally, but ineffably valuable, was lost. Perhaps even the image of the cultured gentleman, and gentlewoman, abhorring the excesses of scientific specialisms, is as worthy an ideal as any. But whatever the merits of these long since passed phenomena, they cannot be conjured up as alternatives for contemporary science policies.

Is then the entire notion of an open community of scholars hopelessly anachronistic? Is full acceptance of the social implications of useful science the only course of action open? Not necessarily. The Stockholm symposium and the contributions to this volume indicate, albeit in a highly tentative fashion, three main responses to the challange of preserving something of the intellectual ethos of open research while taking account of the actual and potential uses of science by the powers-that-be.

A first such response can be termed *the progressive prescription: collective rationality in the face of the future*. Perhaps the Baconian vision of noble, yet useful, men of science is no longer feasible in the age of transient mission-oriented project groups. But could not the enlightened pursuit of a better world which is embodied in the Baconian ideal of 'Salomon's House' be transferred from contemporary 'condottiere' to the enlightened collective of men and women at large, scientists and non-scientists? Could not workable societal arrangements be devised which could accommodate the best features of democratic participation and the aristocratic detachment of the fellows of the 'House of Salomon'? Maybe and maybe not. The confidence in rational design and planning was part and parcel of 'technological sweetness' which has now turned sour. If we can no longer take the collective rationality of scientific communities for granted, what faith can we have in the potential of a much more encompassing collective venture involving all of society in restoring the noble aims of scientific openness and responsibility? What is then called for is a post-Baconian design as imaginative, even visionary, as the one which blazed a trail for the advent of modern science into the yet-to-emerge scientific-industrial society.

165

Many participants in the Stockholm symposium, however, seemed to converge towards a considerably less audacious policy stance which might be labelled *rational policy-making and management: the technological imperative*. In this persepctive, renewed efforts at university-industry interaction have highlighted the fact that the role of universities must be a slightly different one in a high technology society. The continued vitality of university research in such a society depends upon whether both universities and industry recognize that an intensified and mutually beneficial cooperation can be established in which 'free academic research can be preserved in an environment of increasing technocracy' (Hermann Grimmeiss in this volume).

Another version of the rational management position is advocated by Jean-Jacques Salomon at the level of national policy-making. Technological innovation is an absolute prerequisite for renewed economic expansion, and this calls for coherent and long-range oriented research policies which can withstand short-term considerations of budgetary squeeze. 'Coherent strategy-making' must aim at strengthening 'an efficient scientific and technical research system not only in the universities but also, and above all, in industry and the service sector'. Nor can there be any 'question of keeping action on the scientific infrastructure separate from measures to promote the transfer of knowledge and its application to the economic and social system as a whole'.

This is no doubt a highly ambitious prescription. Few would take issue with its objectives, but many might question its practicality. It presupposes that new relationships are established between policy-makers and 'scientists, engineers, technicians and industrialists on the one hand, and trade unions, consumer organizations and representatives of the public on the other' so as to guarantee 'widespread social sanction and commitment' to the aim of restoring higher rates of technical change and economic growth. Although the possibilities of creating or strengthening such relationships should not be under-estimated, it still seems to be a highly open question whether and how such relationships can be shaped to correspond to coherent and encompassing strategic policy-making and planning. After all, there is a wealth of empirical studies and theoretical analyses which suggest that rationalistic long-range planning is liable to run up against any number of difficulties. If the object of planning is something so inherently complex and multifarious as research, technology and higher education and if the number of groupings and interests involved increase dramatically, the prospects for any kind of coordinated and rationalistic planning exercise should be rapidly diminishing.[6]

In her analysis of the role of science for economic growth, Emma Rothschild clearly brings out the inadequacy of more global approaches and shows that many of the crucial problems 'deny the simplicity of macroeconomic policies both in their strictly economic form and as they affect the role of science in economic development'. Hence, it is essential to recognize 'the sectoral and the human complexities of economic growth'. This stance does not necessarily entail abandonment of efforts to formulate well-founded policies for science and technology, but it does caution against bold and broad strategic plans which have too often turned out to be little more than window-dressings or ritualistic evocations of rationality and foresight and, often enough, a recipe for disillusionment in the wake of dashed hopes.

Finally, then, a case might be made for a conception of science *beyond usefulness and overhead, a humanistic response to the challenges to science policy.* Maybe it is not so anachronistic to embrace, to quote Emma Rothschild, the notion that 'the free scientific community is something that is noble in itself' and science as 'something that should be supported in and of itself'. Maybe a conception of science which went beyond traditional views of returns from science 'could actually turn out to be economically useful'. The potentials of a truly innovative search for higher levels of well-being might crucially depend on whether a trust in science as a both disciplined and joyful activity is restored. At the very least, a pluralistic science policy could be promoted which did allow for scholarly activities irrespective of their capacity, no matter how indirectly, to contribute to what has been politically defined as useful and desirable. Such a modest version of a humanistic response should be far from unattainable. It should be fully palatable for the enlightened advocates of science as a necessary overhead. And even in the age of budgetary retrenchments, this latter category still includes most Western science policy-makers.

REFERENCES

1. *Oxford University Gazette*, Supplement (2) to No 3952, 2 April 1984, p 608.
2. *The Times Higher Education Supplement*, August 27 1982.
3. Scott, Peter 1984 *The Crisis of the University*, Croom Helm, London.
4. See Rothblatt, Sheldon 1982 Failure in Early Nineteenth-Century Oxford and Cambridge *History of Education*, **11**: No 1: 1–21.

5. See Wittrock, Björn 1982 Social Knowledge, Public Policy and Social Betterment: A review of Current Research on Knowledge Utilization in Policy-Making. In *European Journal of Political Research* **10**: No 1: 83–89; also Husén, Torsten and Kogan, Maurice (eds) 1984 *Educational Research and Policy: How do they Relate?* Oxford: Pergamon Press.
6. See Clark, Burton R. 1983 *The Higher Education System: Academic Organization in Cross-National Perspective*. Berkeley, Los Angeles and London: University of California Press; Wittrock, Björn 1982 Managing Uncertainty or Foreclosing the Options. In *European Journal of Education* **17**: No 3: 307–318; and 1984 Excellence of Analysis to Diversity of Advocacy. In *Higher Education* **13**: No 2: 121–138.

List of Contributors

Michael Gibbons is Professor and Head of the Department of Science and Technology Policy at Manchester University.

Hermann Grimmeiss is Professor of Physics at the University of Lund and has held leading positions in the electronics industry.

Helga Nowotny is Director of the European Centre for Social Welfare Training and Research, Vienna, and a former Fellow of the Institute for Advanced Study, Berlin.

Sheldon Rothblatt is Professor of History and Chairman of the Department of History at the University of California, Berkeley.

Emma Rothschild is Associate Professor of Technology, Society and Rhetoric, at the Department of Humanities and the College of Science, Technology and Society, MIT.

Jean-Jacques Salomon is Professor of Science Policy at the Conservatoire National des Arts et Métiers, Paris and was for several years Head of the Science Policy Division of the OECD.

Björn Wittrock is Associate Professor of Political Science and Chairman of the Group for the Study of Higher Education and Research Policy at the University of Stockholm.

Dorothy Zinberg is Lecturer at the Center for Science and International Affairs, the Kennedy School of Government, Harvard University.

Index

Index